"In the Working Guides to Estimating and Forecasting Alan has managed to capture the full spectrum of relevant topics with simple explanations, practical examples and academic rigor, while injecting humour into the narrative."

— *Dale Shermon*, Chairman, Society of Cost Analysis and Forecasting (SCAF)

"If estimating has always baffled you, this innovative well illustrated and user friendly book will prove a revelation to its mysteries. To confidently forecast, minimise risk and reduce uncertainty we need full disclosure into the science and art of estimating. Thankfully, and at long last the "Working Guides to Estimating & Forecasting" are exactly that, full of practical examples giving clarity, understanding and validity to the techniques. These are comprehensive step by step guides in understanding the principles of estimating using experientially based models to analyse the most appropriate, repeatable, transparent and credible outcomes. Each of the five volumes affords a valuable tool for both corporate reference and an outstanding practical resource for the teaching and training of this elusive and complex subject. I wish I had access to such a thorough reference when I started in this discipline over 15 years ago, I am looking forward to adding this to my library and using it with my team."

— *Tracey L Clavell*, Head of Estimating & Pricing, BAE Systems Australia

"At last, a comprehensive compendium on these engineering math subjects, essential to both the new and established "cost engineer"! As expected the subjects are presented with the author's usual wit and humour on complex and daunting "mathematically challenging" subjects. As a professional trainer within the MOD Cost Engineering community trying to embed this into my students, I will be recommending this series of books as essential bedtime reading."

— *Steve Baker*, Senior Cost Engineer, DE&S MOD

"Alan has been a highly regarded member of the Cost Estimating and forecasting profession for several years. He is well known for an ability to reduce difficult topics and cost estimating methods down to something that is easily digested. As a master of this communication he would most often be found providing training across the cost estimating and forecasting tools and at all levels of expertise. With this 5-volume set, *Working Guides to Estimating and Forecasting*, Alan has brought his normal verbal training method into a written form. Within their covers Alan steers away from the usual dry academic script into establishing an almost 1:1 relationship with the reader. For my money a recommendable read for all levels of the Cost Estimating and forecasting profession and those who simply want to understand what is in the 'blackbox' just a bit more."

— *Prof Robert Mills*, Margin Engineering, Birmingham City University. MACOSTE, SCAF, ICEAA

"Finally, a book to fill the gap in cost estimating and forecasting! Although other publications exist in this field, they tend to be light on detail whilst also failing to cover many of the essential aspects of estimating and forecasting. Jones covers all this and more from both a theoretical and practical point of view, regularly drawing on his considerable experience in the defence industry to provide many practical examples to support his comments. Heavily illustrated throughout, and often presented in a humorous fashion, this is a must read for those who want to understand the importance of cost estimating within the broader field of project management."

– *Dr Paul Blackwell*, Lecturer in Management of Projects,
The University of Manchester, UK

"Alan Jones provides a useful guidebook and navigation aid for those entering the field of estimating as well as an overview for more experienced practitioners. His humorous asides supplement a thorough explanation of techniques to liven up and illuminate an area which has little attention in the literature, yet is the basis of robust project planning and successful delivery. Alan's talent for explaining the complicated science and art of estimating in practical terms is testament to his knowledge of the subject and to his experience in teaching and training."

– *Therese Lawlor-Wright*, Principal Lecturer in Project Management at the
University of Cumbria

"Alan Jones has created an in depth guide to estimating and forecasting that I have not seen historically. Anyone wishing to improve their awareness in this field should read this and learn from the best."

– *Richard Robinson*, Technical Principal for Estimating, Mott MacDonald

"The book series of 'Working Guides to Estimating and Forecasting' is an essential read for students, academics and practitioners who interested in developing a good understanding of cost estimating and forecasting from real-life perspectives."

– *Professor Essam Shehab*, Professor of Digital Manufacturing and
Head of Cost Engineering, Cranfield University, UK

"In creating the *Working Guides to Estimating and Forecasting*, Alan has captured the core approaches and techniques required to deliver robust and reliable estimates in a single series. Some of the concepts can be challenging, however, Alan has delivered them to the reader in a very accessible way that supports lifelong learning. Whether you are an apprentice, academic or a seasoned professional, these working guides will enhance your ability to understand the alternative approaches to generating a well-executed, defensible estimate, increasing your ability to support competitive advantage in your organisation."

– *Professor Andrew Langridge*, Royal Academy of Engineering
Visiting Professor in Whole Life Cost Engineering and
Cost Data Management, University of Bath, UK

"Alan Jones's "*Working Guides to Estimating and Forecasting*" provides an excellent guide for all levels of cost estimators from the new to the highly experienced. Not only does he cover the underpinning good practice for the field, his books will take you on a journey from cost estimating basics through to how estimating should be used in manufacturing the future – reflecting on a whole life cycle approach. He has written a must-read book for anyone starting cost estimating as well as for those who have been doing estimates for years. Read this book and learn from one of the best."

– *Linda Newnes*, Professor of Cost Engineering, University of Bath, UK

Risk, Opportunity, Uncertainty and Other Random Models

Risk, Opportunity, Uncertainty and Other Random Models (Volume V in the Working Guides to Estimating & Forecasting series) goes part way to debunking the myth that research and development cost are somewhat random, as under certain conditions they can be observed to follow a pattern of behaviour referred to as a Norden-Rayleigh Curve, which unfortunately has to be truncated to stop the myth from becoming a reality! However, there is a practical alternative in relation to a particular form of PERT-Beta Curve.

However, the major emphasis of this volume is the use of Monte Carlo Simulation as a general technique for narrowing down potential outcomes of multiple interacting variables or cost drivers. Perhaps the most common of these is in the evaluation of Risk, Opportunity and Uncertainty. The trouble is that many Monte Carlo Simulation tools are 'black boxes' and too few estimators and forecasters really appreciate what is happening inside the 'black box'. This volume aims to resolve that and to offer tips on things that might need to be considered to remove some of the uninformed random input that often creates a misinformed misconception of 'it must be right!'

Monte Carlo Simulation can be used to model variables that determine Critical Paths in a schedule, and is key to the modelling of Waiting Times and Queues with random arrivals. Supported by a wealth of figures and tables, this is a valuable resource for estimators, engineers, accountants, project risk specialists as well as students of cost engineering.

Alan R. Jones is Principal Consultant at Estimata Limited, an estimating consultancy service. He is a Certified Cost Estimator/Analyst (US) and Certified Cost Engineer (CCE) (UK). Prior to setting up his own business he has enjoyed a 40-year career in the UK aerospace and defence industry as an estimator, culminating in the role of Chief Estimator at BAE Systems. Alan is a Fellow of the Association of Cost Engineers and a Member of the International Cost Estimating and Analysis Association. Historically (some four decades ago), Alan was a graduate in Mathematics from Imperial College of Science and Technology in London, and was an MBA Prize-winner at the Henley Management College (... that was slightly more recent, being only two decades ago). Oh, how time flies when you are enjoying yourself.

Working Guides to Estimating & Forecasting

Alan R. Jones

As engineering and construction projects get bigger, more ambitious and increasingly complex, the ability of organisations to work with realistic estimates of cost, risk or schedule has become fundamental. Working with estimates requires technical and mathematical skills from the estimator but it also requires an understanding of the processes, the constraints and the context by those making investment and planning decisions. You can only forecast the future with confidence if you understand the limitations of your forecast.

The Working Guides to Estimating & Forecasting introduce, explain and illustrate the variety and breadth of numerical techniques and models that are commonly used to build estimates. Alan Jones defines the formulae that underpin many of the techniques; offers justification and explanations for those whose job it is to interpret the estimates; advice on pitfalls and shortcomings; and worked examples. These are often tabular in form to allow you to reproduce the examples in Microsoft Excel. Graphical or pictorial figures are also frequently used to draw attention to particular points as the author advocates that you should always draw a picture before and after analysing data.

The five volumes in the Series provide expert applied advice for estimators, engineers, accountants, project risk specialists as well as students of cost engineering, based on the author's thirty-something years' experience as an estimator, project planner and controller.

Volume I Principles, Process and Practice of Professional Number Juggling
Alan R. Jones

Volume II Probability, Statistics and Other Frightening Stuff
Alan R. Jones

Volume III Best Fit Lines and Curves, and Some Mathe-Magical Transformations
Alan R. Jones

Volume IV Learning, Unlearning and Re-learning Curves
Alan R. Jones

Volume V Risk, Opportunity, Uncertainty and Other Random Models
Alan R. Jones

Risk, Opportunity, Uncertainty and Other Random Models

Alan R. Jones

Routledge
Taylor & Francis Group

LONDON AND NEW YORK

First published in paperback 2024

First published 2019
by Routledge
4 Park Square, Milton Park, Abingdon, Oxon OX14 4RN

and by Routledge
605 Third Avenue, New York, NY 10158

Routledge is an imprint of the Taylor & Francis Group, an informa business

Publisher's Note
The publisher has gone to great lengths to ensure the quality of this reprint but points out that some imperfections in the original copies may be apparent.

British Library Cataloguing-in-Publication Data
A catalogue record for this book is available from the British Library

Library of Congress Cataloging-in-Publication Data
A catalog record has been requested for this book

ISBN: 978-1-138-06505-5 (hbk)
ISBN: 978-1-03-283879-3 (pbk)
ISBN: 978-1-315-16003-0 (ebk)

DOI: 10.4324/9781315160030

Typeset in Bembo
by Apex CoVantage, LLC

To my family:
Lynda, Martin, Gareth and Karl
Thank you for your support and forbearance, and for understanding why
I wanted to do this.

My thanks also to my friends and former colleagues at BAE Systems and the wider
Estimating Community for allowing me the opportunity to learn, develop and practice
my profession ... and for suffering my brand of humour over the years.
In particular, a special thanks to Tracey C, Mike C, Mick P and Andy L for your
support, encouragement and wise counsel. (You know who you are!)

Contents

Figures

Tables

Foreword to the *Working Guides to Estimating and Forecasting* series

At long last a book that will support you throughput your career as an estimator and any other career where you need to manipulate, analyse and more importantly make decisions using your results. Do not be concerned at the book consisting of five volumes as the book is organised into five distinct sections. Whether you are an absolute beginner or an experienced estimator, there will be something for you in these books!

Volume One provides the reader with the core underpinning good-practice required when estimating. Many books miss the need for auditability of your process, clarity of your approach and the techniques you have used. Here, Alan Jones guides you on presenting the basis of your estimate, ensuring you can justify your decisions, evidence these and most of all ensure you keep the focus and understand and focus on the purpose of the model. By the end of this volume you will know how to use e.g. factors and ratios to support data normalisation and how to evidence qualitative judgement. The next volume then leads you through the realm of probability and statistics. This will be useful for Undergraduate students through to experienced professional engineers. The purpose of Volume Two is to ensure the reader '*understands*' the techniques they will be using as well as identifying whether the relationships are statistically significant. By the end of this volume you will be able to analyse data, use the appropriate statistical techniques and be able to determine whether a data point is an outlier or not. Alan then leads us into methods to assist us in presenting non-linear relationships as linear relationships. He presents examples and illustrations for single linear relationships to multi-linear dimensions. Here you do need to have a grasp of the mathematics and the examples and key points highlighted throughout the volumes ensure you can. By the end of this volume you will really grasp best-fit lines and curves.

vAfter Volume Three the focus moves to other influences on your estimates. Volume Four brings out the concept of learning curves – as well as unlearning curves! Throughout this volume you will start with the science behind learning curves but unlike other books, you will get the whole picture. What happens across shared projects and learning, what happens if you have a break in production and have to restart learning. This volume covers the breadth of scenarios that may occur and more importantly how to build these into your estimation process. In my view covering the various types of learning and reflecting these back to real life scenarios is the big win. As stated, many authors focus on learning curves and assume a certain pattern of behaviour. Alan provides you with options, explains these and guides you on how to use them.

The final volume tackles risk and uncertainty. Naturally Monte Carlo simulation is introduced and a guide on really understanding what you are doing. Some of the real winners here are clear hints on guidance on good practice and what to avoid doing. To finalise the book, Alan reflects on the future of Manufacturing where this encompasses the whole life cycle. From his background in Aerospace he can demonstrate the need for critical path in design, manufacture and support along with schedule risk. By considering uncertainty in combination with queueing theory, especially in the spares and repairs domain, Alan demonstrates how the build-up of knowledge from the five volumes can be used to estimate and optimise the whole lifecycle costs of a product and combined services.

I have been waiting for this book to be published for a while and I am grateful for all the work Alan has undertaken to provide what I believe to be a seminal piece of work on the mathematical techniques and methods required to become a great cost estimator. My advice would be for every University Library and every cost estimating team (and beyond) to buy this book. It will serve you through your whole career.

Linda Newnes
Professor of Cost Engineering
Department of Mechanical Engineering
University of Bath
BA2 7AY

1

Introduction and objectives

This series of books aspires to be a practical reference guide to a range of numerical techniques and models that an estimator might wish to consider in analysing historical data in order to forecast the future. Many of the examples and techniques discussed relate to cost estimating in some way, as the term estimator is frequently used synonymously to mean cost estimator. However, many of these numerical or quantitative techniques can be applied in other areas other than cost where estimating is required, such as scheduling, or to determine a forecast of physical, such as weight, length or some other technical parameter.

This volume is a little bit of a mixed bag, but essentially consists mainly of modelling with some known statistical distributions including a liberal dose of estimating with random numbers. (*Yes, that's what a lot of project managers say.*)

1.1 Why write this book? Who might find it useful? Why five volumes?

1.1.1 Why write this series? Who might find it useful?

The intended audience is quite broad, ranging from the relative 'novice' who is embarking on a career as a professional estimator, to those already seasoned in the science and dark arts of estimating. Somewhere between these two extremes of experience, there will be some who just want to know what tips and techniques they can use, to those who really want to understand the theory of why some things work and other things don't. As a consequence, the style of this book is aimed to attract and provide signposts to both (and all those in between).

This series of books is not just aimed at cost estimators (although there is a natural bias there.) There may be some useful tips and techniques for other number jugglers, in which we might include other professionals like engineers or accountants who estimate but do not consider themselves to be estimators *per se*. Also, in using the term 'estimator', we should not constrain our thinking to those whose Estimate's output currency is cost or hours, but also those who estimate in different 'currencies', such as time and physical dimensions or some other technical characteristics.

Finally, in the process of writing this series of guides, it has been a personal voyage of discovery, cathartic even, reminding me of some of the things I once knew but seem to have forgotten or mislaid somewhere along the way. Also, in researching the content, I have discovered many things that I didn't know and now wish I had known years ago when I started on my career, having fallen into it, rather than chosen it (*does that sound familiar to other estimators?*)

1.1.2 Why five volumes?

There are two reasons:

> Size ... there was too much material for the single printed volume that was originally planned ... *and that might have made it too much of a heavy reading so to speak.* That brings out another point, the attempt at humour will remain around that level throughout.
> Cost ... even if it had been produced as a single volume (printed or electronic), the cost may have proved to be prohibitive without a mortgage, and the project would then have been unviable.

So, a decision was made to offer it as a set of five volumes, such that each volume could be purchased and read independently of the others. There is cross-referencing between the volumes, just in case any of us want to dig a little deeper, but by and large the fives volumes can be read independently of each other. There is a common Glossary of Terms across the five volumes which covers terminology that is defined and assumed throughout. This was considered to be essential in setting the right context, as there are many different interpretations of some words in common use in estimating circles. Regrettably, there is a lack of common understanding by what these terms mean, so the glossary clarifies what is meant in this series of volumes.

1.2 Features you'll find in this book and others in this series

People's appetites for practical knowledge varies from the 'How do I?' to the 'Why does that work?' This book will attempt to cater for all tastes.

Many text books are written quite formally, using the third person which can give a feeling of remoteness. In this book, the style used is in first person plural, 'we' and 'us'. Hopefully this will give the sense that this is a journey on which we are embarking together, and that you, the reader, are not alone, especially when it gets to the tricky bits! On that point, let's look at some of the features in this series of *Working Guides to Estimating & Forecasting* ...

1.2.1 Chapter context

Perhaps unsurprisingly, each chapter commences with a very short dialogue about what we are trying to achieve or the purpose of that chapter, and sometimes we might include an outline of a scenario or problem we are trying to address.

1.2.2 The lighter side (humour)

There are some who think that an estimator with a sense of humour is an oxymoron. (*Not true, it's what keeps us sane.*) Experience gleaned from developing and delivering training for estimators has highlighted that people learn better if they are enjoying themselves. We will discover little 'asides' here and there, sometimes at random but usually in italics, to try and keep the attention levels up. (*You're not falling asleep already, are you?*) In other cases the humour, sometimes visual, is used as an *aide memoire*. Those of us who were hoping for a high level of razor-sharp wit, should prepare themselves for a level of disappointment!

1.2.3 Quotations

Here we take the old adage '*A Word to the Wise …*' and give it a slight twist so that we can draw on the wisdom of those far wiser and more experienced in life than I. We call these little interjections '*A word (or two) from the wise*'. You will spot them easily by the rounded shadow boxes. Applying the wisdom of Confucius to estimating, we should explore alternative solutions to achieving our goals rather than automatically de-scoping what we are trying to achieve. The corollary to this is that we should not artificially change our estimate to fit the solution if the solution is unaffordable.

> **A word (or two) from the wise?**
>
> '*When it is obvious that the goals cannot be reached, don't adjust the goals, adjust the action steps.*'
> **Confucius**
> Chinese Philosopher
> 551–479 BC

1.2.4 Definitions

Estimating is not just about numbers but requires the context of an estimate to be expressed in words. There are some words that have very precise meanings; there are others that mean different things to different people (estimators often fall into this latter group). To avoid confusion, we proffer definitions of key words and phrases so that we have a common understanding within the confines of this series of working guides. Where possible we have highlighted where we think that words may be interpreted differently in some sectors, which regrettably, is all too often. I am under no illusion that back in the safety of the real world we will continue to refer to them as they are understood in those sectors, areas and environments.

As the title suggests, this volume is about random models with or without random events, so let's look at what we mean by the term 'random'.

Which do we mean in the context of this volume? Answer: both.

Definition 1.1   Random

Random is an adjective that relates to something that:

1) occurs by chance without method, conscious decision or intervention.
2) behaves in a manner that is unexpected, unusual or odd.

I dare say that some of the definitions given may be controversial with some of us. However, the important point is that they are discussed and considered, and understood in the context of this book, so that everyone accessing these books have the same interpretation; we don't have to agree with the ones given here forevermore – what estimator ever did that? The key point here is that we are able to appreciate that not everyone has the same interpretation of these terms. In some cases, we will defer to the Oxford English Dictionary (Stevenson & Waite, 2011) as the arbiter.

1.2.5 *Discussions and explanations with a mathematical slant for Formula-philes*

These sections are where we define the formulae that underpin many of the techniques in this book. They are boxed off with a header to warn off the faint hearted. We will, within reason, provide justification for the definitions and techniques used. For example:

For the Formula-philes: Exponential inter-arrivals are equivalent to Poisson arrivals

Consider a Poisson Distribution and an Exponential Distribution with a common scale parameter of λ,

Probability of C customer arrivals $p(C)$ in a time period, t, is given by the Poisson Distribution:

$$p(C) = \frac{\lambda^C}{C!} e^{-\lambda} \qquad (1)$$

The Mean of a Poisson Distribution: is its parameter,

Average Customer Arrivals $= \lambda \qquad (2)$

Probability of the Inter-Arrival Time $f(t)$ being t is given by the Exponential Distribution:

$$f(t) = \lambda e^{-\lambda t} \qquad (3)$$

The Mean of an Exponential Distribution is the reciprocal of its parameter, λ

Average InterArrival Time $= \dfrac{1}{\lambda} \qquad (4)$

From (2) and (4) we can conclude that on average:

InterArrival Time $= \dfrac{1}{\text{Customer Arrivals}} \qquad (5)$

… which shows that the Average Inter-Arrival Time in a Time Period is given by the reciprocal of the Average Number of Customer Arrivals in that Time Period

1.2.6 Discussions and explanations without a mathematical slant for Formula-phobes

For those less geeky than me, who don't get a buzz from knowing why a formula works (*yes, it's true, there are some estimators like that*), there are the Formula-phobe sections with a suitable header to give you more of a warm comforting feeling. These are usually wordier with pictorial justifications, and with specific particular examples where it helps the understanding and acceptance.

For the Formula-phobes: One-way logic is like a dead lobster

An analogy I remember coming across reading as a fledgling teenage mathematician, but for which sadly I can no longer recall its creator, relates to the fate of lobsters. It has stuck with me, and I recreate it here with my respects to whoever taught it to me.

Sad though it may be to talk of the untimely death of crustaceans, the truth is that all boiled lobsters are dead! However, we cannot say that the reverse is true – not all dead lobsters have been boiled!

(Continued)

One-way logic is a response to many-to-one relationship in which there are many circumstances that lead to a single outcome, but from that outcome we cannot stipulate what was the circumstance that led to it.

Please note that no real lobsters were harmed in the making of this analogy.

1.2.7 Caveat augur

Based on the fairly well-known warning to shoppers: '*Caveat Emptor*' (let the buyer beware) these call-out sections provide warnings to all soothsayers (or estimators) who try to predict the future, that in some circumstances we many encounter difficulties in using some of the techniques. They should not be considered to be foolproof or be a panacea to cure all ills.

Caveat augur

These are warnings to the estimator that there are certain limitations, pitfalls or tripwires in the use or interpretation of some of the techniques. We cannot profess to cover every particular aspect, but where they come to mind these gentle warnings are shared

1.2.8 Worked examples

There is a proliferation of examples of the numerical techniques in action. These are often tabular in form to allow us to reproduce the examples in Microsoft Excel (*other spreadsheet tools are available*). Graphical or pictorial figures are also used frequently to draw attention to particular points. The book advocates that we should '**always draw a picture before and after analysing data.**' In some cases, we show situations where a particular technique is unsuitable (i.e. it doesn't work) and try to explain why. Sometimes we learn from our mistakes; nothing and no-one is infallible in the wondrous world of estimating. The tabular examples follow the spirit and intent of Best Practice Spreadsheet Modelling (albeit limited to black and white in the absences of affordable colour printing), the principles and virtues of which are summarised here in Volume I Chapter 3.

1.2.9 Useful Microsoft Excel functions and facilities

Embedded in many of the examples are some of the many useful special functions and facilities found within Microsoft Excel (*often, but not always, the estimator's toolset of choice because of its flexibility and accessibility*). Together we explore how we can exploit these functions and features in using the techniques described in this book.

We will always provide the full syntax as we recommend that we avoid allowing Microsoft Excel to use its default settings for certain parameters when they are not specified. This avoids unexpected and unintended results in modelling and improves transparency, an important concept that we discussed in Volume I Chapter 3.

Example:

> The **SUMIF(*range*, *criteria*, *sum_range*)** function will summate the values in the *sum_range* if the *criteria* in *range* is satisfied, and exclude other values from the sum where the condition is not met. Note that *sum_range* is an optional parameter of the function in Excel; if it is not specified then the *range* will be assumed instead. We recommend that we specify it even if it is the same. This is not because we don't trust Excel, but a person interpreting our model may not be aware that a default has been assumed without our being by their side to explain it.

1.2.10 References to authoritative sources

Every estimate requires a documented Basis of Estimate. In common with that principle, which we discussed in Volume I Chapter 3, every chapter will provide a reference source for researchers, technical authors, writers and those of a curious disposition, where an original, more authoritative or more detailed source of information can be found on particular aspects or topics.

Note that an Estimate without a Basis of Estimate becomes a random number in the future. On the same basis, without reference to an authoritative source, prior research or empirical observation becomes little more than a spurious unsubstantiated comment.

1.2.11 Chapter reviews

Perhaps not unexpectedly, each chapter summarises the key topics that we will have discussed on our journey. Where appropriate we may draw a conclusion or two just to bring things together, or to draw out a key message that may run throughout the chapter.

1.3 Overview of chapters in this volume

Volume V begins in earnest with a discussion in Chapter 2 on how we can model the research and development, concept demonstration or design and development tasks when we only know the objectives of the development, but not how we are going to achieve them in detail. One possible solution may be to explore the use of a Norden-Rayleigh Curve, which essentially is a repeating pattern of resource and cost consumption over time that has been shown empirically to follow the natural pattern of problem discovery and resolution over the life of such 'solution development' projects. However, using a Norden-Rayleigh Curve is not without its pitfalls and is not a panacea for this type of work. Consequently, we will explore some alternative options in the shape (as it were) of Beta, PERT-Beta, Triangular and Weibull Distributions.

Chapter 3 is based fundamentally on the need to derive 3-Point Estimates; we discuss how we can use Monte Carlo Simulation to model and analyse Risk, Opportunity and Uncertainty variation. As Monte Carlo Simulation software is generally proprietary in nature, and is often under-understood by its users, we discuss some of the 'do's and don'ts' in the context of Risk, Opportunity and Uncertainty Modelling, not least of which is how and when to apply correlation between apparently random events! However, Monte Carlo Simulation is not a technique that is the sole reserve of the Risk Managers and the like; it can also be used to test other assumptions in a more general modelling and estimating sense. We conclude the chapter with the warning that the use of a Bottom-up Monte Carlo Simulation for Risk, Opportunity and Uncertainty analysis is fundamentally optimistic, a weakness that is often overlooked.

To overcome this weakness, we take a more holistic view of Risk Opportunity and Uncertainty in Chapter 4 by taking a top-down perspective and exploring the use of the 'Marching Army' technique and Uplift Factors to give us a more pessimistic view of our estimate. We offer up the Slipping and Sliding Technique as a simple way of combining the Top-down and Bottom-up perspectives to get a more realistic view of the likely range of outcomes. Finally, we throw Estimate and Schedule Maturity Assessments back into the mix that we introduced in Volume I Chapter 3, as a means of gauging the maturity of our final estimate.

In Chapter 5 we warn about the dangers of the much-used Risk Factoring or Expected Value Technique. We discuss and demonstrate how it is often innocently but ignorantly abused, to quantify risk contingency budgets. We discuss better ways to do this using a variation on the same simple concept.

The need for planning resurfaces in Chapter 6 with an introduction to Critical Path Analysis and discuss how this can be linked with Monte Carlo Simulation and Correlation to perform Schedule Risk Analysis. This chapter concentrates on how we can determine the Critical Path using a simple technique with Binary Numbers.

In the last chapter of this final volume we discuss Queueing Theory (*it just had to be last, didn't it?*) and how we might use it in support of achievable solutions where we have random arisings (such as spares or repairs) against which we need to develop a viable estimate. We discuss what we mean by a Memoryless System and how with Monte Carlo Simulation we can generate meaningful solutions or options.

1.4 Elsewhere in the 'Working Guide to Estimating & Forecasting' series

Whilst every effort has been made to keep each volume independent of others in the series, this would have been impossible without some major duplication and overlap.

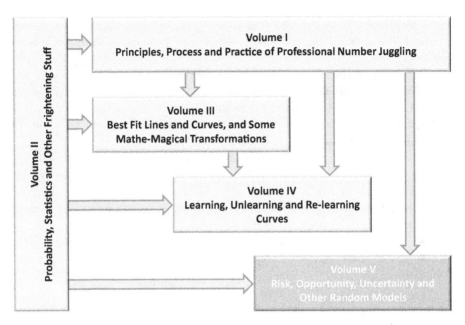

Figure 1.1 Principal Flow of Prior Topic Knowledge Between Volumes

Whilst there is quite a lot of cross-referral to other volumes, this is largely for those of us who want to explore particular topics in more depth. There are some more fundamental potential pre-requisites. For example, in relation to the evaluation of Risks, Opportunities and Uncertainties discussed in this volume, the relevance of 3-Point Estimates was covered in Volume I, and a more thorough discussion of the probability and statistics that underpin Monte Carlo Simulation can be found in Volume II.

Figure 1.1 indicates the principal linkages or flows across the five volumes, not all of them.

1.4.1 Volume I: Principles, Process and Practice of Professional Number Juggling

This volume clarifies the differences in what we mean by an Estimating Approach, Method or Technique, and how these can be incorporated into a closed-loop Estimating Process. We discuss the importance of TRACEability and the need for a well-documented Basis of Estimate that differentiates an estimate from what would appear in the future to be little more than a random number. Closely associated with a Basis of Estimate is the concept of an Estimate Maturity Assessment, which in effect gives us a Health Warning on the robustness of the estimate that has been developed. IRiS is a companion tool that allows us to assess the inherent risk in our estimating spreadsheets and models if we fail to follow good practice principles in designing and compiling those spreadsheets or models.

An underlying theme we introduce here is the difference between accuracy and precision within the estimate, and the need to check how sensitive our estimates are to changes in assumptions. We go on to discuss how we can use factors, rates and ratios in support of Data Normalisation (to allow like-for-like comparisons to be made) and in developing simple estimates using an Analogical Method.

All estimating basically requires some degree of quantitative analysis, but we will find that there will be times when a more qualitative judgement may be required to arrive at a numerical value. However, in the spirit of TRACEability, we should strive to express or record such subjective judgements in a more quantitative way. To aid this we discuss a few pseudo-quantitative techniques of this nature.

Finally, to round off this volume, we will explore how we might use Benford's Law, normally used in fraud detection, to highlight potential anomalies in third party inputs to our Estimating Process.

1.4.2 Volume II: Probability, Statistics and Other Frightening Stuff

Volume II is focused on the Statistical concepts that are exploited through Volumes III to V (and to a lesser extent in Volume I). It is not always necessary to read the associated detail in this volume if you are happy just to accept and use the various concepts, principles and conclusions. However, a general understanding is always better than blind acceptance, and this volume is geared around making these statistical topics more accessible and understandable to those who wish to adventure into the darker art and science of estimating. There are also some useful 'Rules of Thumb' that may be helpful to estimators or other number jugglers that are not directly used by other volumes.

We explore the differences between the different statistics that are collectively referred to as 'Measures of Central Tendency' and why they are referred to as such. In this discussion, we consider four different types of Mean (Arithmetic, Geometric, Harmonic and Quadratic) in addition to Modes and the 'one and only' Median, all of which are, or might be, used by estimators, sometimes without our conscious awareness.

However, the Measures of Central Tendency only tell us half the story about our data, and we should really understand the extent of scatter around the Measures of Central Tendency that we use; this gives us valuable insight to the sensitivity and robustness of our estimate based on the chosen 'central value'. This is where the 'Measures of Dispersion and Shape' come into their own. These measures include various ways of quantifying the 'average' deviation around the Arithmetic Mean or Median, as well as how we might recognise 'skewness' (where data is asymmetric or lop-sided in its distribution), and where our data exhibits high levels of Excess Kurtosis, which measures how spikey our data scatter is relative to the absolute range of scatter. The greater the Excess Kurtosis, and the more symmetrical our data, then the greater confidence we should have in the Measures of Central Tendency being representative of the majority of our data. Talking of 'confidence' this leads us to explore Confidence Intervals and Quantiles, which are frequently used to describe the robustness of an estimate in quantitative terms.

Extending this further we also explore several probability distributions that may describe the potential variation in the data underpinning our estimates more completely. We consider a number of key properties of each that we can exploit, often as 'Rules of Thumb', but that are often accurate enough without being precise.

Estimating in principle is based on the concept of Correlation, which expresses the extent to which the value of one 'thing' varies with another, the value of which we know or have assumed. This volume considers how we can measure the degree of correlation, what it means and, importantly, what it does not mean! It also looks at the problem of a system of variables that are partially correlated, and how we might impose that relationship in a multi-variate model.

Estimating is not just about making calculations, it requires judgement, not least of which is whether an estimating relationship is credible and supportable, or 'statistically significant'. We discuss the use of Hypothesis Testing to support an informed decision when making these judgement calls. This approach leads naturally onto Tests for 'Outliers'. Knowing when and where we can safely and legitimately exclude what looks like unrepresentative or rogue data from our thoughts is always a tricky dilemma for estimators. We wrap up this volume by exploring several Statistical Tests that allow us to 'Out the Outliers'; be warned however, these various Outlier tests do not always give us the same advice!

1.4.3 Volume III: Best Fit Lines and Curves, and Some Mathe-Magical Transformations

This volume concentrates on fitting the 'Best Fit' Line or Curve through our data, and creating estimates through interpolation or extrapolation and expressing the confidence we have in those estimates based on the degree of scatter around the 'Best Fit' Line or Curve.

We start this volume off quite gently by exploring the properties of a straight line that we can exploit, including perhaps a surprising non-linear property. We follow this by looking at simple data smoothing techniques using a range of 'Moving Measures' and stick a proverbial toe in the undulating waters of exponential smoothing. All these techniques can help us to judge whether we do in fact have an underlying trend that is either linear (straight line) or non-linear (curved).

We begin our exploration of the delights of Least Squares Regression by considering how and why it works with simple straight-line relationships before extending it out into additional 'multi-linear' dimensions with several independent variables, each of which is linearly correlated with our dependent variable that we want to estimate. A very important aspect of formal Regression Analysis is measuring whether the Regression relationship is credible and supportable.

Such is the world of estimating, that many estimating relationships are not linear, but there are three groups of relationships (or functions) that can be converted into linear relationships with a bit of simple mathe-magical transformation. These are Exponential, Logarithmic and Power Functions; some of us will have seen these as different Trendline types in Microsoft Excel.

We then demonstrate how we can use this mathe-magical transformation to convert a non-linear relationship into a linear one to which we can subsequently exploit the power of Least Squares Regression.

Where we have data that cannot be transformed in to a simple or multi-linear form, we explore the options open to us to find the 'Best Fit' curve, using Least Squares from first principles, and exploiting the power of Microsoft Excel's Solver.

Last, but not least, we look at Time Series Analysis techniques in which we consider a repeating seasonal and/or cyclical variation in our data over time around an underlying trend.

1.4.4 Volume IV: Learning, Unlearning and Re-Learning Curves

Where we have recurring or repeating activities that exhibit a progressive reduction in cost, time or effort we might want to consider Learning Curves, which have been shown empirically to work in many different sectors.

We start our exploration by considering the basic principles of a learning curve and the alternative models that are available, which are almost always based on Crawford's Unit Learning Curve or the original Wright's Cumulative Average Learning Curve. Later in the volume we will discuss the lesser used Time-based Learning Curves and how they differ from Unit-based Learning Curves. This is followed by a healthy debate on the drivers of learning, and how this gave rise to the Segmentation Approach to Unit Learning.

One of the most difficult scenarios to quantify is the negative impact of breaks in continuity, causing what we might term Unlearning or Forgetting, and subsequent Re-learning. We discuss options for how these can be addressed in a number of ways, including the Segmentation Approach and the Anderlohr Technique.

There is perhaps a misconception that Unit-based Learning means that we can only update our Learning Curve analysis when each successive unit is completed. This is not so, and we show how we can use Equivalent Units Completed to give us an 'early warning indicator' of changes in the underlying Unit-based Learning.

We then turn our attention to shared learning across similar products or variants of a base product through Multi-Variant Learning, before extending the principles of the segmentation technique to a more general transfer of learning across between different products using common business processes.

Although it is perhaps a somewhat tenuous link, this is where we explore the issue of Collaborative Projects in which work is shared between partners, often internationally with workshare being driven by their respective national authority customers based on their investment proportions. This generally adds cost due to duplication of effort and an increase in integration activity. There are a couple of models that may help us to estimate such impacts, one of which bears an uncanny resemblance to a Cumulative Average Learning Curve (*I said that it was a tenuous link.*)

1.4.5 Volume V: Risk, Opportunity, Uncertainty and Other Random Models

This is where we are now. This section is included here just to make sure that the paragraph numbering aligns with the volume numbers! (*Estimators like structure; it's engrained; we can't help it.*)

We covered this in more detail in Section 1.3, so we will not repeat or summarise it further here.

1.5 Final thoughts and musings on this volume and series

In this chapter, we have outlined the contents of this volume and to some degree the others in this series, and described the key features that have been included to ease our journey through the various techniques and concepts discussed. We have also discussed the broad outline of each chapter of this volume, and reviewed an overview of the other volumes in the series to whet our appetites. We have also highlighted many of the features that are used throughout the five volumes that comprise this series, to guide our journey and hopefully make it less painful or traumatic.

One of the key themes in this volume is how estimators can use random number simulation models such as Monte Carlo to estimate and forecast costs and schedules. Some of the main uses of these techniques include the generation of 3-Point Estimates for Risk, Opportunity and Uncertainty, for both cost and schedule. The same techniques can be used to develop models underpinned by Queueing Theory, which help establish operation models and support Service Availability Contracts.

However, the danger of toolsets which offer bespoke Monte Carlo Simulation capability, is that a 'black-box' mentality can set in, and estimators, forecasters and schedulers often plug numbers in at one end and extract them from the other without really understanding what is happening by default in-between. One of the intents of this volume is to take that 'black-box' and give it more of a more transparent lid, so that we better understand what is going on inside!

A word (or two) from the wise?

'No matter how many times the results of experiments agree with some theory, you can never be sure that the next time the result will not contradict the theory.'

Stephen Hawking
British Physicist
b.1942

However, we must not delude ourselves into thinking that if we follow these techniques slavishly that we won't still get it wrong some of the time, often because assumptions have changed or were misplaced, or we made a judgement call that perhaps we wouldn't have made in hindsight. A recurring theme throughout this volume, and others in the series, is that it is essential that we document what we have done and why; TRACE-ability is paramount. The techniques in this series are here to help guide our judgement through an informed decision-making process, and to remove the need to resort to 'guesswork' as much as possible.

As Stephen Hawking (1988, p.11) reminds us just because a model appears to work well, is no guarantee that it is correct, or that we won't get freak results!

TRACE: Transparent, Repeatable, Appropriate, Credible and
Experientially-based

References

Hawking, S (1988) *A Brief History of Time*, London, Bantam Press.
Stevenson, A & Waite, M (Eds), (2011) *Concise Oxford English Dictionary (12th Edition)*, Oxford, Oxford University Press.

2 | Norden-Rayleigh Curves for solution development

In Volume IV we discussed ways to model and estimate the time, cost or effort of recurring or repetitive tasks, but we haven't looked at what happens before that ... how long, or how much effort do we require to develop the solution that becomes the basis of our repetitive task, or of developing that one-off task, especially if it is novel or innovative?

Even if we can develop that estimate by analogy to another previous and not too dissimilar task, or by parametric means, there is nothing obvious so far that helps us say how that estimate might be profiled in time; a problem that is often more of a challenge perhaps than getting the base value itself.

Perhaps the best-known model for doing this is the Norden-Rayleigh Curve, but there are others that we might refer to collectively as 'Solution Development Curves'.

2.1 Norden-Rayleigh Curves: Who, what, where, when and why?

John William Strut, the 3rd Lord Rayleigh, was an English physicist who discovered the gas Argon (for which he won a Nobel Prize in 1904); he also described why the sky appears to be blue. (*We could say that he was one of life's true 'Blue Sky Thinkers' who knew how to strut his stuff.*) More importantly for us as estimators, he came up with the Probability Distribution that bears his title, not his name. (*The Rayleigh Distribution has a better ring to it than the Strut Distribution, doesn't it? That would make it sound rather like a component stress calculation.*) The Rayleigh

> ### A word (or two) from the wise?
>
> *'If we knew what it was we were doing, it would not be called research, would it?'*
>
> Attributed to
> **Albert Einstein**
> Physicist
> 1879–1955

Distribution is often used to model reliability and failure rates, but has numerous other applications as well.

In 1963, Peter Norden contemplated the life cycle of a basic development process from a pragmatic rather than a purely theoretical perspective in relation to the software engineering sector. Paraphrasing his deliberations, which we might summarise as 'trial and error', the process of natural development is:

- Someone makes a request for something that doesn't exist
- Someone comes up with a bright idea for a solution
- Resource is deployed to scope the requirements in more detail and identify what they mean, and in so doing creates a list of problems to overcome (*beginning to sound familiar?*)
- More resource is deployed to overcome the problems
- These problems are overcome but throw up more understanding of the needs of the solution but also raises other problems, and perhaps better ways of tackling what has already been done ... all of which need more resources. *(If this sounds like we don't know what we're doing, then just reflect on the thoughts of Albert Einstein (attributed by Amneus & Hawken, 1999, p.272).)*
- Eventually the combined resources break the back of the mound of problems, and less and less new challenges are discovered, and existing solutions can be used to complete the remaining tasks, which ultimately need fewer resources
- Eventually, the solution is developed (although often the engineers will probably have thought of some enhancements and have to be forcibly prised away from the project (*a crowbar is a useful project management tool in this instance!*)

By analysing the profile of many such development curves, Norden concluded that a truncated Rayleigh Distribution was a good fit. The Rayleigh Distribution appeared to perform well as a model for the ramp-up of resources to a peak before tapering off again over time. The only problem is that a Rayleigh Probability Distribution extends to positive infinity, and so from a pragmatic point of view, it has to be truncated. By convention (*implying that this is accepted practice, rather than something that is absolutely correct*), this is taken to be at the 97% Confidence Level of a Rayleigh Distribution (Lee, 2002). However, this does cause some issues, as we will explore later, and there may be some practical alternatives we can consider instead. In other words, there is no compelling reason why we cannot have more than one convention, just as we have with physical measurement scales (Imperial vs. Metric).

Although its roots lie in the project management of software engineering projects, further studies by David Lee and others (e.g. Gallagher & Lee, 1996) have shown that the basic principles of the Norden-Rayleigh Curve can be extended to more general research and development tasks. In particular, it has been shown to fit some major defense development programmes (Lee, Hogue & Hoffman, 1993). This seems to indicate that there is an underlying consistency and commonality across complex engineering development processes. This latter article demonstrated that the cost and schedule of such development programmes could be normalised in terms of their relative scales into a single curve – the Norden-Rayleigh Curve. The implication for us, therefore, is that (theoretically speaking) we can stretch or compress the overall curve without losing its basic shape or integrity. It implies also that there are two scale factors, one for each of the horizontal and vertical scales. However, from a more empirical perspective, this does not necessarily work once we have commenced the development process.

2.1.1 Probability Density Function and Cumulative Distribution Function

Although the Norden-Rayleigh Curve is based on the Rayleigh Distribution, the Probability Density Function and Cumulative Distribution Function are not the most elegant to read or follow.

The good news is that the Rayleigh Distribution is actually a special case of the Weibull Distribution in which the Weibull parameters $\alpha = 2$ (*always*) and $\beta = \lambda\sqrt{2}$, where λ is the Rayleigh Distribution Mode parameter. Why is this good news? As we discussed in Volume II Chapter 4 on Probability Distributions (*unless you skipped that particular delight and left me talking to myself*) Microsoft Excel has a special in-built function for the Weibull Distribution:

WEIBULL.DIST(*x, alpha, beta, cumulative*)

- To calculate either the probability density value for x (*cumulative* = FALSE), or the Cumulative Probability (*cumulative* = TRUE) with Shape Parameter, **alpha** and Scale Parameter **beta**
  In earlier versions of Microsoft Excel, the function was simply:
  WEIBULL(*x, alpha, beta, cumulative*)

For the Formula-philes: Rayleigh Distribution

For a Rayleigh Distribution with parameter, λ:

Probability Density Function (PDF), $f(x)$:

$$f(x) = \frac{x}{\lambda^2} e^{-\frac{1}{2}\left(\frac{x}{\lambda}\right)^2}$$

Cumulative Distribution

Function (CDF), $F(x)$:

$$F(x) = 1 - e^{-\frac{1}{2}\left(\frac{x}{\lambda}\right)^2}$$

(Continued)

Key Parameters/Statistics		Value
Measures of Central Tendency	Mean	$\lambda\sqrt{\dfrac{\pi}{2}}$
	Median	$\lambda\sqrt{2\ln 2}$
	Mode	λ
Measures of Dispersion	Minimum	0
	Maximum	∞
	Variance	$\lambda^2\left(2-\dfrac{\pi}{2}\right)$

So, a Rayleigh Distribution would be **WEIBULL.DIST(*x*, 2, *Mode* $\sqrt{2}$, *TRUE or FALSE*)** in Microsoft Excel depending on whether we want the CDF or PDF values.

To convert this into a Norden-Rayleigh Curve to represent a time-phased profile of the development cost, we simply express the independent variable, x to be time from the project start, multiplied by the estimate of the total cost … and then decide where we are going to truncate it.

2.1.2 Truncation options

As the distribution is scalable in the sense that we can stretch or compress it vertically to change the project outturn values, or we can stretch or compress it horizontally to reflect a change in the chosen rate of development, i.e. its schedule, then we only need to examine the relative properties against a 'standard distribution'. For this purpose, we will only consider the cost profile values (vertical axis) in relation to the percentage of the final outturn cost being 100%. For the horizontal axis, we will look at time elapsed from the start relative to a multiplier of the mode position. For simplification, we will define the mode to occur at time 1 relative to a start time of 0. Table 2.1 and Figure 2.3 illustrates the invariable percentage values from the Cumulative Distribution Function.

As we will observe, the Mode occurs when Development is theoretically 39.347% complete (*that'll be at around 40% in Estimating Speak!*).

The 'conventional' truncation point for a Norden-Rayleigh Curve is at the 97% Confidence Level (Lee, 2002), which always occurs at 2.65 times the Mode. (*We might as well call it two and two-thirds.*)

However, for those of us who are uncomfortable with this, being perhaps a somewhat premature truncation point, we will see that there are other Rules of Thumb we can use that infer a truncation at different levels of decimal place precision closer to 100%, such as 3.5 times the Mode. The nice thing about the 3.5 factor is that it is visually close

Table 2.1 Cumulative Distribution Function of a Standard Rayleigh Probability Distribution

Time	Rayleigh Distribution	Comment
0	0.000%	
0.25	3.077%	
0.5	11.750%	
0.75	24.516%	
1	39.347%	< Mode
1.25	54.217%	
1.41	63.212%	< Mode * SQRT(2)
1.5	67.535%	
1.75	78.373%	
2	86.466%	
2.25	92.044%	
2.5	95.606%	
2.65	97.014%	< Conventional N-R Truncation Point
2.75	97.721%	
2.83	98.168%	
3	98.889%	
3.25	99.491%	
3.5	99.781%	< 100% to 0 dp's
3.75	99.912%	
4	99.966%	< 100% to 1 dp
4.25	99.988%	
4.5	99.996%	< 100% to 2 dp's
4.75	99.999%	
5	100.000%	< 100% to 3 dp's

enough to 100% (see Figure 2.1) and it is also analogous to the Normal Distribution's 'standard Confidence Interval' of the Mean ±3 Standard Deviations (99.73%).

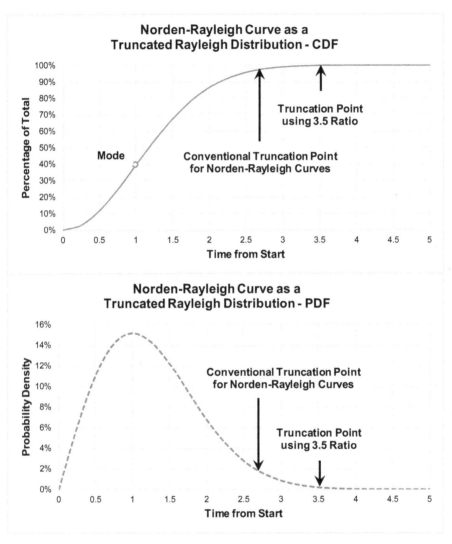

Figure 2.1 Norden-Rayleigh Curve Compared with a Standard Rayleigh Probability Distribution

The question we should be asking ourselves when we select a truncation point is:

How precisely inaccurate do we want our profile to be?

Some of us may be wondering why we have highlighted the 63.212% at the Mode x SQRT(2) in Table 2.1. This links back to the standard property of a Weibull Distribution (Volume II Chapter 4 if you missed that particular delight).

There are three benefits to be had with choosing an end-point at the 3.5 factor point in time:

i. We don't 'have to' factor the vertical scale to accommodate the truncation because if we do choose to do so, the Uplift Factor is only 1.0022. (*Our development estimate is not going to be that precise, is it?*)

ii. The PDF can be used to model the resource profile by time period without a sudden cut-off implied by the conventional approach (see Figure 2.1.) However, if this is what happens in reality, then perhaps we should adopt the convention of a 2.65 Truncation Ratio unswervingly at the 97% Completion Level.

iii. We don't have a long and procrastinated tail to higher levels of hypothetical precision for true values greater than 3.5 times the Mode.

There is a case to be made that a value of 3 is an acceptable compromise, achieving a level of some 98.9% with a much easier multiplier to remember and without the additional crawl through to 3.5 for a paltry 0.9%. (*Try using that argument with an accountant when the total is in the millions or billions!*)

Any truncation activity should consider what is likely to happen in the environment in question. Ask ourselves whether any development is truly completed in an absolute sense. If the answer is 'yes' then in terms of recurring production learning there would be no contribution required from engineering etc. So at what point do we just say 'enough is enough' and terminate the development (*probably to the annoyance or relief of some engineers*)?

2.1.3 How does a Norden-Rayleigh Curve differ from the Rayleigh Distribution?

The main difference between the two is one of scale. The cumulative of any probability distribution is always 100% (or 1) whereas the Norden-Rayleigh Curve is geared to the final outturn cost. This is complicated because the Rayleigh Distribution as a special case of a Weibull Distribution 'goes on indefinitely' and by implication only reaches 100% at infinity. Design and Development programmes do not go on indefinitely (*although sometimes, to some of us, some of them appear to go on forever*) and a practical limitation has to be placed on them, hence the manual truncation of the cost-schedule curve.

For the Formula-philes: Norden-Rayleigh Curve

Consider a Norden-Rayleigh Curve with an outturn value of C_T (the vertical scale parameter) and a completion time of T relative to its start time at zero, and a mode at time λ (the horizontal scale parameter.) Let C_∞ be the notional cost at infinity if development was allowed to continue indefinitely.

Let the Cumulative Cost at time t on a Norden-Rayleigh Curve with Mode λ be denoted as $N_\lambda(t)$:

$$N_\lambda(t) = C_\infty \left(1 - e^{-\frac{1}{2}\left(\frac{t}{\lambda}\right)^2} \right) \tag{1}$$

At time T the value C_T is achieved:

$$C_T = N_\lambda(T) = C_\infty \left(1 - e^{-\frac{1}{2}\left(\frac{T}{\lambda}\right)^2} \right) \tag{2}$$

Substituting C_∞ from (2) in (1):

$$N_\lambda(t) = \frac{C_T}{\left(1 - e^{-\frac{1}{2}\left(\frac{T}{\lambda}\right)^2} \right)} \left(1 - e^{-\frac{1}{2}\left(\frac{t}{\lambda}\right)^2} \right) \tag{3}$$

Let k be a constant (>1) uplift factor based on the Completion time, T such that:

$$k = \frac{1}{\left(1 - e^{-\frac{1}{2}\left(\frac{T}{\lambda}\right)^2} \right)} \tag{4}$$

Substituting (4) in (3):

$$N_\lambda(t) = kC_T \left(1 - e^{-\frac{1}{2}\left(\frac{t}{\lambda}\right)^2} \right) \tag{5}$$

In Figure 2.2, we can see the characteristic asymmetric S-Curve of a Cumulative Rayleigh Distribution Function (CDF) with its positive skew. In comparison, we can see how the Norden-Rayleigh Curve truncates and uplifts the Rayleigh Distribution. In this example we have truncated and uplifted the Rayleigh Distribution at the 97% Confidence/Completion Level.

Note that we have not stipulated any definition with regards to the scale of the Time axis. That's because we can define it in relation to any scale, be it days, weeks, lunar months, calendar months or years; *it can also be in hours, minutes, seconds, decades, centuries etc. but these are totally impractical which is why I never mentioned them ... oh! I just have now ... well please ignore them.*

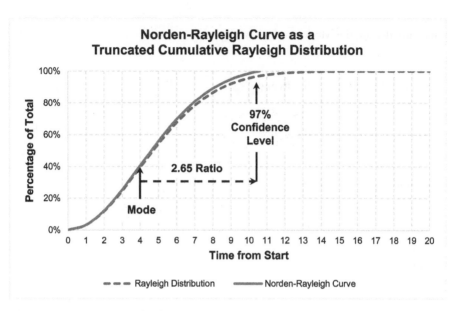

Figure 2.2 Norden–Rayleigh Curve as a Truncated Cumulative Rayleigh Probability Distribution

For the Formula-philes: Time is a relative term with a Norden-Rayleigh Curve

Consider a Norden-Rayleigh Curve with an outturn value of C_T (the vertical scale parameter) and a completion time of T relative to a start time of zero, and a mode at time λ.

Let the Cumulative Cost at time t on a Norden-Rayleigh Curve with

Mode λ be denoted as $N_\lambda(t)$:

$$N_\lambda(t) = \frac{C_T}{\left(1 - e^{-\frac{1}{2}\left(\frac{T}{\lambda}\right)^2}\right)}\left(1 - e^{-\frac{1}{2}\left(\frac{t}{\lambda}\right)^2}\right) \quad (1)$$

Let's convert the time to a new scale denoted by $\tilde{t}$ such that for a constant α:

$$t = \alpha\tilde{t} \quad (2)$$

Similarly denote the Completion time on the new scale by $\tilde{T}$ such that:

$$T = \alpha\tilde{T} \quad (3)$$

(Continued)

Similarly denote the Mode on the new scale as occurring at time $\tilde{T}$ where:

$$\lambda = \alpha \tilde{\lambda} \qquad (4)$$

Dividing (2) by (4) and simplifying:

$$\frac{t}{\lambda} = \frac{\tilde{t}}{\tilde{\lambda}} \qquad (5)$$

Dividing (3) by (4) and simplifying:

$$\frac{T}{\lambda} = \frac{\tilde{T}}{\tilde{\lambda}} \qquad (6)$$

Substituting (5) and (6) in (1):

$$N_\lambda(t) = \frac{C_T}{\left(1 - e^{-\frac{1}{2}\left(\frac{\tilde{T}}{\tilde{\lambda}}\right)^2}\right)} \left(1 - e^{-\frac{1}{2}\left(\frac{\tilde{t}}{\tilde{\lambda}}\right)^2}\right) \qquad (7)$$

But by definition the Norden-Rayleigh Curve with Mode $\tilde{\lambda}$ be denoted as $N_{\tilde{\lambda}}(t)$:

$$N_{\tilde{\lambda}}(t) = \frac{C_T}{\left(1 - e^{-\frac{1}{2}\left(\frac{\tilde{T}}{\tilde{\lambda}}\right)^2}\right)} \left(1 - e^{-\frac{1}{2}\left(\frac{\tilde{t}}{\tilde{\lambda}}\right)^2}\right) \qquad (8)$$

Therefore, from (7) and (8):

$$N_\lambda(t) = N_{\tilde{\lambda}}(\tilde{t})$$

… i.e. Norden-Rayleigh Curves are independent of the Time axis measurement scale

We can easily redefine the Time axis measurement scale by applying different scale factors. For instance, the scale factor to convert a NRC expressed in years to one expressed in months is 12. Table 2.2 illustrates that there is no difference. Both curves use the same Excel function for the Rayleigh Distribution but use the different Time scales.

The left hand Curve uses WEIBULL.DIST(year, 2, 1.414, TRUE) where 1.414 is the Mode of 1 multiplied by the Square Root of 2.
The right hand Curve uses WEIBULL.DIST(month, 2, 16.971, TRUE) where 16.971 is the Mode of 12 multiplied by the Square Root of 2

The Norden-Rayleigh Curve equivalent of the Rayleigh Probability Density Function (PDF) is the relative resource loading or cost expenditure 'burn rate' at a point in time. We can see the consequences of the truncation and uplift of the former to the latter in Figure 2.3.

Table 2.2 Norden-Rayleigh Curves are Independent of the Time Scale

	1	year =	12	months
Time	**Rayleigh Distribution**	**Time**	**Rayleigh Distribution**	
Start	0	Start	0	
End	3.50	End	42	
Mode	1	Mode	12	
alpha	2	alpha	2	
beta	1.414	=Mode x √2	beta	16.971	=Mode x √2

	Time (Years)	Rayleigh Distribution CDF	Norden-Rayleigh Uplift	Time (Months)	Rayleigh Distribution CDF	Norden-Rayleigh Uplift	
End	3.50	99.781%	100.219%	42.00	99.781%	100.219%	
		↳ Reciprocal ↱			↳ Reciprocal ↱		

Time (Years)	Rayleigh Distribution	Norden-Rayleigh Curve	Time (Months)	Rayleigh Distribution	Norden-Rayleigh Curve	Rayleigh Distribution x Uplift
0	0.000%	0.000%	0	0.000%	0.000%	
0.5	11.750%	11.776%	6	11.750%	11.776%	
1	39.347%	39.433%	12	39.347%	39.433%	
1.5	67.535%	67.683%	18	67.535%	67.683%	
2	86.466%	86.656%	24	86.466%	86.656%	
2.5	95.606%	95.816%	30	95.606%	95.816%	
3	98.889%	99.106%	36	98.889%	99.106%	
3.5	99.781%	100.000%	42	99.781%	100.000%	

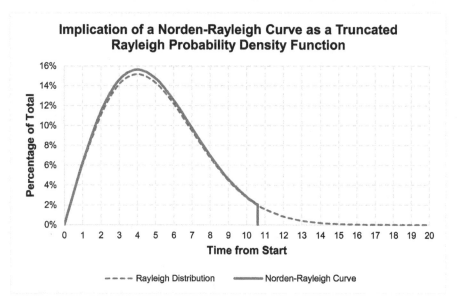

Implication of a Norden-Rayleigh Curve as a Truncated Rayleigh Probability Density Function

- - - - Rayleigh Distribution ▬▬▬ Norden-Rayleigh Curve

Figure 2.3 Implication of a Norden-Rayleigh Curve as a Truncated Rayleigh Probability Density Function

2.1.4 Some practical limitations of the Norden-Rayleigh Curve

Dukovich et al (1999) conducted an extensive study into the practical application of Norden-Rayleigh Curves and expanded our understanding of how, when and why they are appropriate in the research and development environment:

- The characteristic shape and constancy of the Rayleigh Distribution implies that there is a relationship between the resource loading and the total effort or cost expended in achieving the development goals.
- Any development project has a finite number of problems which must be overcome but not all these problems are known at the start of the project.
- Most of the development cost incurred must be the labour cost incurred by the development team in the time they take to resolve problems. Non-labour costs should be relatively small. (It is probably safe for us to assume in many instances that any costs for outsourced development is considered to be predominantly labour costs.)
- The total number of people actively engaged on the development at a given time is in directly proportion to the number of problems being worked at that time, and their relative complexity (Norden, 1963).
- In resolving problems, this uncovers other problems which were previously unforeseen. This effectively limits the introduction of new members of the team (and their skill requirements) until the problems are identified. As a consequence, it is often impractical to compress development timescales except in cases where 'resource starvation' is a matter of policy, e.g. where it is temporarily unavailable due to other priorities. If additional resource (people) are added prematurely, this adds cost but achieves virtually nothing as there is a risk that they will be underutilised.

A word (or two) from the wise?

Augustine's Law of Economic Unipolarity:

'The only thing more costly than stretching the schedule of an established project is accelerating it, which is itself the most costly action known to man.'

Norman R Augustine
Law No XXIV
1997

As Augustine (1997, p.158) rightly points out, tinkering with a schedule once it has been established is going to be expensive.

Dukovich et al (1999) concurred with Putnam & Myers (1992) that the development project must also be a bounded development project through to its completion with a consistent set of objectives, i.e. no change in direction. However, they did consider the aspect of concurrent or phased development projects and we will return to this later in Section 2.2.1.

In essence, the development objective is expected to remain largely (if not totally) unchanged, and there is no realistic expectation that spend profiles on incremental development projects, or ones whose objectives evolve over time will be adequately represented by a Norden-Rayleigh Curve. Likewise, any project that we undertake that has a finite resource cap which is less than that required to meet the emerging problems to be solved, may not follow the pattern of the Norden-Rayleigh Curve. However, again we will revisit that particular scenario later.

In some respects, it may seem to be very limiting and that a Norden-Rayleigh Curve might only be applicable to small, relatively simple, research and development projects. Not so! Using data published by the RAND Corporation (Younossi et al, 2005), we can demonstrate that they work for major Aircraft Engineering & Manufacturing Development (EMD) programmes such as the United States F/A-18 and F/A-22 Fighter Aircraft (*except they would have spelt it with one 'm' not two*).

In fact, the normalised expenditure profile is a reasonable approximation to a Rayleigh Distribution for F/A-18E/F, heralded as a successful development programme in terms of cost and schedule adherence (Younossi et al, 2005), as we can see from Figure 2.4. The Actual Cost of Work Performed data (ACWP) has been extracted with permission from

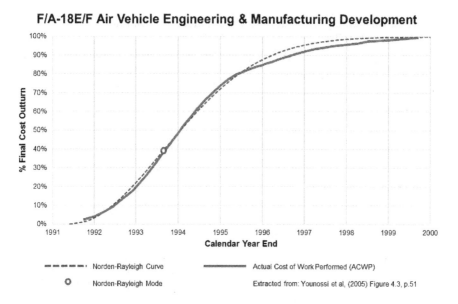

Figure 2.4 Indicative Example of F/A-18 EMD Costs Plotted Against a Rayleigh Distribution

Source: © 2005 RAND Corporation

Younossi O, Stem DE, Lorell MA & Lussier FM, (2005) Lessons Learned from the F/A-22 and F/A-18 E/F Development Programs, Santa Monica, CA, RAND Corporation.

Figure 4.3 of the RAND Report (Younossi et al, 2005, p.51) and superimposed on a Norden-Rayleigh Curve; the tail-end is less well-matched but it does recover in the latter stages. We might speculate that this could be as a consequence of funding constraints in the latter stages.

In respect of the F/A-22 as shown in Figure 2.5, the ACWP data (extracted from Younossi et al (2005) Figure 4.1, p.39) is a very good fit to the Norden-Rayleigh Curve … at least as far as the data published allows us; we will note that it is incomplete with potentially an additional 7% cost to be incurred to complete the development programme. It is possible that this spend profile was subsequently truncated. We will revisit the implications of this discussion in Section 2.5.

We might want to reflect on this. Military Aircraft Development Projects are rarely 'single, integrated development projects' with unchanging objectives as allegedly required for a Norden-Rayleigh to be valid, but they seem to work well here … and potentially without the need for the 'conventional' truncation at 97%. Here we have used the 3.5 Truncation Ratio at the 99.97% Confidence Level.

Putnam and Myers (1992) had previously applied a variation of the Norden-Rayleigh Curve to the development of software applications. These are often referred to as the Putnam-Norden-Rayleigh Curve (or PNR Curves) and date back to 1976.

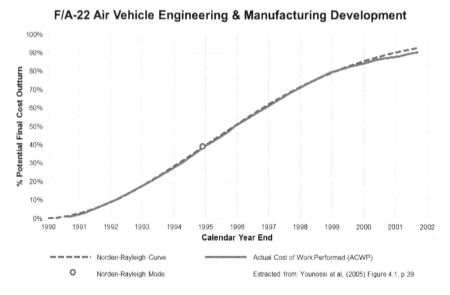

Figure 2.5 Indicative Example of F/A-22 EMD Costs Plotted Against a Rayleigh Distribution

Source: © 2005 RAND Corporation

Younossi O, Stem DE, Lorell MA & Lussier FM, (2005) Lessons Learned from the F/A-22 and F/A-18 E/F Development Programs, Santa Monica, CA, RAND Corporation.

However, there are many instances (Dukovich et al, 1999) where the principles of a Norden-Rayleigh Curve are not met and hence the Rayleigh Distribution is not (or may not) be appropriate:

a) **Software Development** projects characterised by:
   - A phased development and release of interim issues
   - Evolving or changing development objectives following each interim release
   - An emphasis on 'bug fixing' and minor enhancements to existing versions (Yes, yes, I can hear the cynics muttering 'Well, that'll be all software projects then!')

b) **Phased Development** projects/contracts in which:
   - The objectives of each development phase are not clearly distinct from each other
   - There is a gap between the phases, and the resourcing strategy differs for each phase

Note: where the resourcing strategy is the same, the gap may be accommodated by '*stopping the clock*' for the duration of the gap, or slowing it down during the ramp down of one phase and the ramp-up of the next. We will look at this more closely in Section 2.2.4.

c) **Independent Third-Party Dependencies** (*sounds like an oxymoron but isn't*) in which:
   - The development requires an activity to be performed that is not in the control of the development project but is essential to the natural ongoing development process. This could include Government Testing using dedicated facilities for which there are other independent demands and constraints on its availability
   - There is another parallel development programme (with its own objectives) on which there is an expectation of a satisfactory outcome to be exploited by the current project
   - The delays caused by independent third-parties lead to additional resource ramp-downs, the inevitable attrition of key members of the team and subsequent ramp-up again after the delay ... possibly with new team members

Note: It may be possible to accommodate such delays by '*stopping the clock*' or slowing it down for the duration of the delay if this occurs during the ramp-up or ramp-down phases of the project rather than during the main problem resolution phase. Again, we will look at this more closely in Section 2.2.4.

d) **Multiple Prototype Development Programmes** in which:
   - Each prototype is a variation of the first and there is a high degree of commonality requiring the purchase of many common items
   - There are multiple hardware purchases, requiring little or no development investment, but which are disproportionate to the scale of the development tasks interspersed between the purchase required

- The procurement activity and delivery is often phased to attract economically favourable terms and conditions from suppliers, creating spikes and plateaus in expenditure profiles
- The unique development task peculiar to each new prototype variant may be shown to follow the pattern of a Norden–Rayleigh Curve if resource, time and cost are recorded separately for each prototype, but where the same resource is working on multiple tasks this may 'distort' the profile of every prototype

In all cases, especially where the development extends across a number of years it is strongly recommended that all costs and labour time are normalised to take account of economic differences such as escalation, and any adjustments to accounting conventions including changes in overhead allocations (see Volume I Chapter 6).

2.2 Breaking the Norden-Rayleigh 'Rules'

Despite the tongue-in-cheek nature of the 'Law of Economic Unipolarity' (Augustine, 1997) to which we referred in Section 2.1.2, stretching the schedule of development programmes is not unknown and as Augustine's 24th Law suggests is not without its consequences. However, let's look at schedule elongation due to four different reasons:

1. Additional work scope (on top of and not instead of the original objectives)
2. Correction of an overly optimistic view of the problem complexity
3. Inability to recruit resource to the project in a timely fashion
4. Premature resource reduction (e.g. due to budget constraint)

2.2.1 Additional objectives: Phased development (or the 'camelling')

In theory, this could be the easiest to deal with. It is not unreasonable to assume that the additional development objectives also follow the Norden-Rayleigh Curve and principles. If the additional work scope is independent of the original work scope other than it may be dependent of the satisfactory resolution of some problem from the original scope in order to begin or progress beyond a certain point. Let's consider the first of these semi-independent conditions.

Suppose we have an additional 20% work scope that we can start when the original project is around two-thirds complete; that would be at Time period 1.5 relative to a mode at time period 1. Figure 2.6 shows the addition of two Norden-Rayleigh Curves with identical relative modes that only differ in their start date and their modal value. The resultant curve, although positively skewed, is not a Norden-Rayleigh Curve as the ratio of the duration to the Mode is 5.

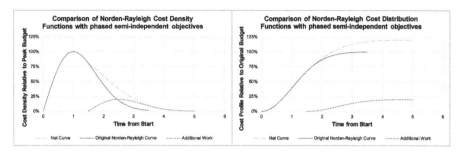

Figure 2.6 Phased Semi-Independent Developments with an Elongated Single Hump

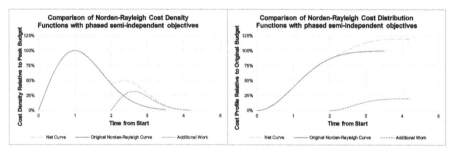

Figure 2.7 Phased Semi-Independent Developments with Suppressed Double Hump

This is something of a fluke because if the start point of the second development was earlier or later, or its duration was shorter, we would most probably get a suppressed double hump as shown in Figure 2.7. So, if the first example was a Dromedary (single hump) Camel, then this is tending towards being more of a Bactrian (double-humped) Camel. The later into the original schedule the additional work is commenced, and the more significant the work content is, then the more Bactrian-like the development resource profile will look.

However, if the additional work scope is more interdependent on the original objectives, then it may be more appropriate to consider this to be more akin to an underestimation of the project complexity …

2.2.2 Correcting an overly optimistic view of the problem complexity: The Square Rule

In this situation, we are considering the situation where perhaps the original estimate or the allocated budget was too low (*Shock, horror! It's true: estimators do get it wrong sometimes, and other times they just get overruled!*)

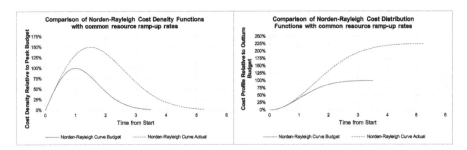

Figure 2.8 The Penalty for Underestimating the Complexity of Solution Development

If the project is being allowed to ramp-up resource naturally as new problems are discovered then the principles will continue to apply, reaching a higher mode at a later time, and eventually finishing later than originally planned. In cases like this, we can use the inviolable shape property of the Norden-Rayleigh Curve to assess the likely cost of schedule slippage. In fact, it's a *Square Rule*; a percentage increase in schedule generates an increase in cost equal to the square of the slippage increase, adding some insight into Augustine's observation that schedule slippage is extremely expensive.

We can illustrate this with Figure 2.8. The NRC Cost Density Functions in the left-hand graph show that the actual mode occurred 50% later than the budgeting assumption with a corresponding knock-on to the project completion date. Also, we have the situation where the actual Modal Value was 50% greater than that assumed in the budget, but that initially the ramp-up rates were the same. In the right-hand graph, we show the corresponding cumulative cost distributions; whilst the development project '*only*' took 50% longer than planned, the outturn cost was a whacking 125% over budget, i.e. 225% is the square of 150%. Such is the potential penalty for underestimating engineering complexity, giving us our **Square Rule**.

This rule also works in cases where the development turns out to be less complex than originally expected and we find that we don't need to ramp-up our resource as high and for as long. In theory, this allows us to shorten the schedule with the consequential *square root* benefit on cost.

However, do not confuse natural early completion with artificial compression …

The same principle of squaring the scale factors can be extended to any distribution-based resource profile and its associated Cumulative S-Curve, as illustrated by the Formula-phobe call-out on the cost of schedule slippage based on a Triangular Distribution.

Caveat augur

Shortening a project schedule because the project is simpler than expected is not the same as compressing the schedule artificially to accelerate the completion date without any corresponding reduction in assumed complexity. The intellectual thought processes and physical validation results synonymous with Solution Development generally cannot be hastened.

For the Formula-phobes: Square Rule for the cost of schedule slippage

If we accept the principles and properties of the Norden-Rayleigh Curve then the effective end date is some 3.5 times the Mode or point of peak resource or expenditure.

The S-Curve is the area under the resource profile graph. So if the resource is proportionately taller and lasts for longer, then the area will increase by the product of the two. If the two proportions are the same then the profile will be increased by the square of the slippage.

Consider a Triangular Distribution as an analogy. (We could liken the Norden-Rayleigh Distribution to be a sort of 'rounded triangle'.)

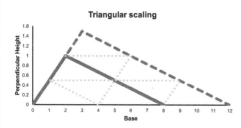

As we may recall from school Maths lessons (*unless those memories are still too painful*), the area of a triangle is always: **'Half the Base times the Perpendicular Height'**

If we scale both the height and the base by the same factor we keep the same shape but change the area by the square of the scaling factor.

In the example, we have scaled the base and height by 1.5 and the area is then 2.25 times bigger. The smaller triangle has an area of 4, whilst the area of the larger dotted triangle with the same basic shape, has an area of 9. Each mini triangle has an area of 1.

For the Formula-philes: Norden-Rayleigh Square Rule

Consider a Norden-Rayleigh Curve (NRC) with an Outturn Cost of C_λ and a Mode at $_\lambda$, based on a truncated Rayleigh Distribution using an uplift factor of k:

Rayleigh Distribution Probability Density Function (PDF), $f(x)$:

$$f(x) = \frac{x}{\lambda^2} e^{-\frac{1}{2}\left(\frac{x}{\lambda}\right)^2} \tag{1}$$

The NRC equivalent factored cost density function $N_\lambda(t)$ with a Mode at λ can be expressed as:

$$N_\lambda(t) = kC_\lambda \left(\frac{t}{\lambda^2} e^{-\frac{1}{2}\left(\frac{t}{\lambda}\right)^2} \right) \tag{2}$$

From (2), when $t = \lambda$, the NRC Modal Value is:

$$N_\lambda(\lambda) = kC_\lambda \left(\frac{\lambda}{\lambda^2} e^{-\frac{1}{2}\left(\frac{\lambda}{\lambda}\right)^2} \right) \tag{3}$$

Simplifying (3):

$$N_\lambda(\lambda) = kC_\lambda \left(\frac{1}{\lambda} e^{-\frac{1}{2}} \right) \tag{4}$$

Taking the ratio of the Modal Value to the Mode, λ as a measure of the NRC resource ramp up rate, R_λ:

$$R_\lambda = \frac{N_\lambda(\lambda)}{\lambda} \tag{5}$$

Substituting (4) in (5) and simplifying:

$$R_\lambda = kC_\lambda \left(\frac{1}{\lambda^2} e^{-\frac{1}{2}} \right) \tag{6}$$

Similarly for a second NRC with a Mode at m_λ, and outturn cost $C_{k\lambda}$ where m is a constant multiplier:

$$R_{k\lambda} = kC_{m\lambda} \left(\frac{1}{m^2\lambda^2} e^{-\frac{1}{2}} \right) \tag{7}$$

From (6) and (7), if the Resource Ramp-up Rates are the same:

$$\frac{C_{m\lambda}}{\tau} \left(\frac{1}{m^2\lambda^2} e^{-\frac{1}{2}} \right) = \frac{C_\lambda}{\tau} \left(\frac{1}{\lambda^2} e^{-\frac{1}{2}} \right) \tag{8}$$

Simplifying and re-arranging (8):

$$C_{m\lambda} = m^2 C_\lambda$$

… if the ratio of the Modal Value and the Mode of two Norden-Rayleigh Cost Density Curves are constant, the outturn Cost will differ by the square of the factor

2.2.3 Schedule slippage due to resource ramp-up delays: The Pro Rata Product Rule

If we accept the principles and properties of the Norden-Rayleigh Curve, then the number of problems to be solved are finite, and many of these problems are unknown until earlier problems are investigated (problems create problems.) If we cannot resource our project quickly enough, we may discover a new problem that has to wait until the resource is available. Let's say that we eventually resource up to the same level of resource as planned, albeit late. We can exploit the shape integrity of the Norden-Rayleigh Curve to estimate the impact of the schedule delay. In this case the cost increase is *pro rata* to the schedule slippage as illustrated in Figure 2.9. Here, the actual Mode occurs 50% later than expected and the overall project also finishes 50% later. In contrast to the previous scenario, we are assuming that the peak resource (Modal Value) remains the same; the net result is that the cost overrun is *pro rata* to the schedule slippage, giving us a **Pro Rata Product Rule**.

Now you might well ask why we have referred to this as 'Pro Rata Product Rule' and not simply the 'Pro Rata Rule'. Well, consider this … in some instances we may not need to resource up to the same level as existing resource may be freed up to look at new problems, thus reducing the need for additional resource. In this case, we may find that there is no impact on cost outturn even though there is a schedule slippage. Let's look at that one in Figure 2.10. In this case even though the schedule slips by 50%, the actual peak resource is only two-thirds of the budgeted level. If we take the product of the Peak Resource factor (i.e. ⅔) and the Schedule Change factor (1.5), we get unity, i.e. our factors cancel out.

I can see that one or two of us have spotted that the previous Square Rule is just a special case of the Pro Rata Product Rule, in which both factors are the same. Maybe we can exploit this in the third and most difficult context, i.e. when budget restrictions are introduced after the project has commenced.

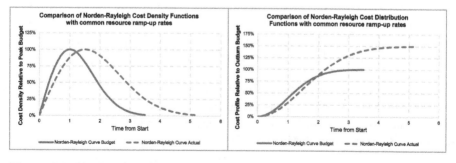

Figure 2.9 The Penalty of Late Resourcing

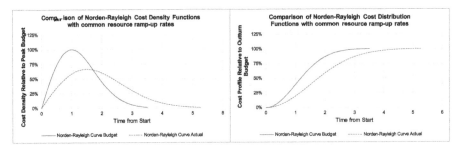

Figure 2.10 The Opportunity of Late Resourcing

2.2.4 Schedule slippage due to premature resource reduction

Here we are considering a development project in which the project has already been started and has been geared to an original schedule, but later, possibly after the peak resource has been passed, funding is restricted going forward, or resource is re-prioritised to another project. In this case, stretching the cycle due to a budgetary constraint disrupts Norden's hypothesis of how and why resource ramps up sharply initially as new problems are discovered, and then decays later over a longer period.

It can be argued that the need to take off resource quickly or prematurely leaves some problems unresolved until the remaining resource is free to take them on board. Potentially this could imply some duplication of prior effort as they review the work done previously by those who have now departed the project (*or worse, they simply just repeat it*). How do we deal with such an eventuality, *or do we just throw our hands in the air and resign ourselves to the fact that they will probably just blame any overspend on the estimator? (We can probably still feel the pain and indignity of it.)*

Let's consider a controlled slow down, where for reasons of cash flow, or alternative business priorities, the development project needs to free up half its resource from a point in time. Inevitably, the schedule will slip, but what will happen to the cost? Let's look at how we might tackle it from a few different perspectives:

1. Slowing Down Time (The Optimistic Perspective)
2. Marching Army Penalty (A More Pragmatic Perspective)
3. Creating a Bactrian Camel Effect (The Disaggregated Perspective)
4. Modal Re-positioning (A Misguided Perspective)

Let's look at each in turn.

(1) Slowing Down Time (The Optimistic Perspective)

We could say that anyone who thinks that they can slow down time is optimistic (or deluded!).

From an optimistic perspective, we could argue that had we known that we would only have access to half the resource in the first place, then we might have assumed that the development cycle would have taken twice as long. We could splice two Norden-Rayleigh Curves together using the fully resourced curve until the resource capping is imposed, followed by the capped resourced curve until the development is completed. In effect this is the same as saying that from the point of the resource reduction to 50% of the previous level, we can achieve the same resolution to our challenges but in twice the time.

The easiest way to create the data table for this is to slow down time! The following procedure explains what we mean by this.

1. Create the normal Norden-Rayleigh Curve for both the Cost Density Function and the Cumulative Distribution Function (as per the Budget) based on a truncated Rayleigh Distribution.
2. Multiply the Cost Density Function by the resource reduction factor (in this example 50%).
3. Create a second Time variable which matches the original time used in Step 2 UNTIL the point at which the artificial spend constraint is to be imposed. After that point increase the time increment by dividing by the resource reduction factor. This in effect slows down time!

See the example in Table 2.3 (showing values until time = 2 for brevity) and Figure 2.11. The left hand graph depicts the resource profile and the right hand graph the cumulative cost profile.

Now we can see why this approach is an optimistic one (*some would say 'naïve'*) in that there is no prime cost penalty, only a schedule penalty, irrespective of when the

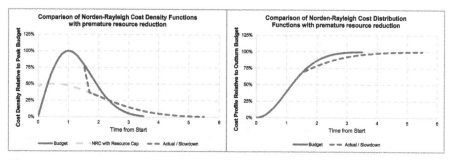

Figure 2.11 Optimistic View of Premature Resource Reduction

Table 2.3 Stretching Time to Model the Optimistic View of Premature Resource Reduction

	Time	Rayleigh Distribution			
	alpha	2			
	beta	1.414		Slowdown Resource Factor	
Norden-Rayleigh Truncation Ratio >	Start	0			
	End	3.50			
	Mode	1		50%	

Budget Time	Rayleigh Distribution CDF	Norden-Rayleigh Uplift	Peak Resource	Slowdown from Time
3.50	99.781%	100.219%	60.786%	1.50

Budget Time	Rayleigh Distribution	Norden-Rayleigh Curve	NRC Resource as % of Peak	Slowdown Resource as % of Peak	Stretched Time
0	0.000%	0.000%	0.000%	0.000%	0.00
0.1	0.499%	0.500%	16.405%	16.405%	0.10
0.2	1.980%	1.984%	32.321%	32.321%	0.20
0.3	4.400%	4.410%	47.285%	47.285%	0.30
0.4	7.688%	7.705%	60.878%	60.878%	0.40
0.5	11.750%	11.776%	72.750%	72.750%	0.50
0.6	16.473%	16.509%	82.628%	82.628%	0.60
0.7	21.730%	21.777%	90.332%	90.332%	0.70
0.8	27.385%	27.445%	95.777%	95.777%	0.80
0.9	33.302%	33.375%	98.969%	98.969%	0.90
1	39.347%	39.433%	100.000%	100.000%	1.00
1.1	45.393%	45.492%	99.036%	99.036%	1.10
1.2	51.325%	51.437%	96.302%	96.302%	1.20
1.3	57.044%	57.169%	92.069%	92.069%	1.30
1.4	62.469%	62.606%	86.630%	86.630%	1.40
1.5	67.535%	67.683%	80.289%	80.289%	1.50
1.6	72.196%	72.355%	73.345%	36.672%	1.70
1.7	76.425%	76.593%	66.076%	33.038%	1.90
1.8	80.210%	80.386%	58.730%	29.365%	2.10
1.9	83.553%	83.736%	51.523%	25.761%	2.30
2	86.466%	86.656%	44.626%	22.313%	2.50
2.1	88.975%	89.170%	38.172%	19.086%	2.70
2.2	91.108%	91.308%	32.254%	16.127%	2.90
2.3	92.899%	93.103%	26.926%	13.463%	3.10
2.4	94.387%	94.593%	22.212%	11.106%	3.30
2.5	95.606%	95.816%	18.110%	9.055%	3.50
2.6	96.595%	96.807%	14.595%	7.298%	3.70
2.7	97.388%	97.601%	11.628%	5.814%	3.90
2.8	98.016%	98.231%	9.159%	4.580%	4.10
2.9	98.508%	98.724%	7.134%	3.567%	4.30
3	98.889%	99.106%	5.495%	2.747%	4.50
3.1	99.181%	99.399%	4.185%	2.093%	4.70
3.2	99.402%	99.620%	3.153%	1.576%	4.90
3.3	99.568%	99.786%	2.349%	1.175%	5.10
3.4	99.691%	99.910%	1.731%	0.866%	5.30
3.5	99.781%	100.000%	1.262%	0.631%	5.50

Time increments increase after Slowdown

intervention occurs! (Any cost penalty will come from additional inflationary aspects due to slippage in time.) The schedule penalty is to double the time remaining to completion.

For many of us, however, that will seem to be counter-intuitive, and certainly conflicts with the observations of Augustine in relation to schedule slippage. The reason that it has happened is that it makes the presumption that there is no handover or rework penalty. It implies that had we started earlier with half the resource then we would have reached the same point in development at the same time. (*Really? I wish you luck with that one!*)

Let's look at the basic principle of it from a more pessimistic viewpoint with a handover or rework penalty.

(2) Marching Army Penalty (A More Pragmatic Perspective)

In terms of a penalty (*conspicuous by its absence in the previous option*) one possible model we might consider adopts the principle of the 'Marching Army' (*some people prefer the term 'Standing Army'*):

1. We follow a standard Norden-Rayleigh Curve until the intervention point (in this case at period 1.5).
2. We reduce the level of resource available to us … in this example to a half.
3. We create another Norden-Rayleigh Curve using the reduced resource and a proportionately longer schedule (in this case double the schedule with half the resource) and align it horizontally so that it intersects the original curve at the position equivalent to the reduced resource level.
4. We maintain a Marching Army of constant resource from the point at which we reduced the resource at step 2 until the slow-down curve from step 3 intersects the original curve from step 1 (as determined in step 3).
5. We then follow the slow-down curve through to completion, gradually reducing the resource.
6. The cumulative version is then the two Norden-Rayleigh Curves spliced together either side of a linear 'slug' corresponding to the duration of the 'Marching Army'.

Figure 2.12 illustrates the effect of this on the resource profile and the resultant cumulative cost profile also, showing a cost penalty of some 13.8%.

The problem we will have with this model and technique is in determining the horizontal offset of the slow-down curve and the intersection point of the two resource profiles. Table 2.4 illustrates how we can use Microsoft Excel's Solver to help us with this procedure.

In this little model we have only one variable … the offset for the second Norden-Rayleigh Curve. The model exploits the property of the Rayleigh Distribution as the key component of the Norden-Rayleigh Curve. If we factor the resource (equivalent to the Probability Density Function) of a curve, we extend the overall distribution cycle by the inverse of the factor; so in this case halving the resource doubles the cycle time.

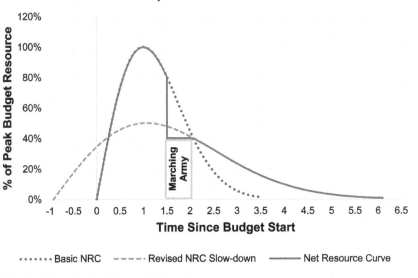

Comparison of Norden-Rayleigh Cost Density Functions with premature resource reduction

Legend: Basic NRC, Revised NRC Slow-down, Net Resource Curve

Comparison of Norden-Rayleigh Cost Distribution Functions with premature resource reduction

Legend: Basic NRC, Revised NRC Slow-down, Net Cost Profile

Figure 2.12 More Pragmatic View of Premature Resource Reduction Using the Marching Army Principle

Table 2.4 Determining the Marching Army Parameters

	Original Rayleigh Distribution		Revised Rayleigh Distribution	
Weibull Alpha Parameter	2		2	
Relative Resource	100%		50%	
Rayleigh Mode	1		2	
Weibull Beta Parameter	1.414	= Mode √2	2.828	= Mode √2
Offset Start Time	0		-0.931	<0 Left Shift, >0 Right Shift
NRC Mode to End Factor	3.5		3.5	
End Time	3.50		6.07	=3.5 x Mode + Offset

Intervention Time	Intersection Time	Rayleigh Distribution PDF	Time from Offset to Intersection	Rayleigh Distribution PDF	PDF Difference (Error)
1.5	2.069	24.349%	3.000	24.349%	0.00%

Intersection Time + Offset
Relative Resource

Table 2.5 Determining the Marching Army Parameters – Alternative Model

	Value	Value Divided by Mode	Identity Required		
			Natural Log	Minus Half Square	Sum of Natural Log & Minus Half Square
Original Mode	1				
Revised Mode Factor, p	2				
Inverse Factor, 1/p	50%		-0.693	#N/A	-1.413	< Values
Intervention Time, T	1.5	1.500	0.405	-1.125		< should
Intersection Time, X	2.069	2.069	0.727	-2.140	-1.413	< match
Offset Time, d = X - pT	-0.931				0.000	< Difference

We can then find the intersection point where the two Probability Density Functions are equal by allowing the second to be offset to the right or left. The degree to which it moves is dependent on the Intervention Time relative to the original curve's start time of zero. The only constraint that we have enforced here is that the Offset must be less than the Mode of the original curve … otherwise we might get a match on the upward side rather than the downward side of the distribution. An alternative but somewhat less intuitive Solver model can be set up as in Table 2.5.

This Solver model relies on the identity in row (8) of the adjacent Formula-phile call-out on 'Determining the Marching Army parameters'. The model allows the Inter-section Time to vary until the Difference between the two calculations based on the Intervention Time and the Intersection Time is zero. Note that formula based on the

Intervention Time also includes the Natural Log of the Modal Factor (the inverse of the resource reduction factor).

For the Formula-philes: Determining the Marching Army parameters

Consider a Rayleigh Distribution with a Scale parameter, λ, which is its Mode. Consider a second Rayleigh Distribution with a Mode at a factor p times that of the first distribution.

Probability Density Function (PDF), $f(t)$ of the first distribution is:

$$f(t) = \frac{t}{\lambda^2} e^{-\frac{1}{2}\left(\frac{t}{\lambda}\right)^2} \qquad (1)$$

At time pt the PDF value, $g(pt)$ of the second distribution is:

$$g(pt) = \frac{pt}{p^2\lambda^2} e^{-\frac{1}{2}\left(\frac{pt}{p\lambda}\right)^2} \qquad (2)$$

Simplifying (2):

$$g(pt) = \frac{t}{p\lambda^2} e^{-\frac{1}{2}\left(\frac{t}{\lambda}\right)^2} \qquad (3)$$

Substituting (1) in (3):

$$g(pt) = \frac{1}{p} f(t) \qquad (4)$$

Let X be that time when the first distribution equals value of $g(pT)$:

$$f(X) = \frac{1}{p} f(T) \qquad (5)$$

Substituting (1) in (5):

$$\frac{X}{\lambda^2} e^{-\frac{1}{2}\left(\frac{X}{\lambda}\right)^2} = \frac{1}{p}\frac{T}{\lambda^2} e^{-\frac{1}{2}\left(\frac{T}{\lambda}\right)^2} \qquad (6)$$

Simplifying (6):

$$pXe^{-\frac{1}{2}\left(\frac{X}{\lambda}\right)^2} = Te^{-\frac{1}{2}\left(\frac{T}{\lambda}\right)^2} \qquad (7)$$

Taking Natural Logs of (7):

$$\ln(p) + \ln(X) - \frac{1}{2}\left(\frac{X}{\lambda}\right)^2 = \ln(T) - \frac{1}{2}\left(\frac{T}{\lambda}\right)^2 \qquad (8)$$

Define δ to be the difference between time X and pT:

$$\delta = X - pT \qquad (9)$$

... from which we can use (8) and (9) to solve the point at which two Rayleigh Distributions intersect, one of which is offset from zero

Table 2.6 Marching Army Parameters and Penalties

Resource Cap	Relative Intervention Point	Cumulative % Completion	Curve Intersection	Start Offset	Estimate At Completion	Nominal End Point
25%	1	39.4%	2.339	-1.661	120.3%	12.34
25%	1.5	67.7%	2.450	-3.550	111.6%	10.45
25%	2	86.7%	2.718	-5.282	104.8%	8.72
25%	2.5	95.8%	3.072	-6.928	101.6%	7.07
25%	3	99.1%	3.474	-8.526	100.4%	5.47
50%	1	39.4%	1.922	-0.078	127.9%	6.92
50%	1.5	67.7%	2.069	-0.931	113.8%	6.07
50%	2	86.7%	2.398	-1.602	105.4%	5.40
50%	2.5	95.8%	2.805	-2.195	101.6%	4.80
50%	3	99.1%	3.247	-2.753	100.4%	4.25
75%	1	39.4%	1.577	0.243	126.0%	4.91
75%	1.5	67.7%	1.780	-0.220	110.2%	4.45
75%	2	86.7%	2.179	-0.488	103.6%	4.18
75%	2.5	95.8%	2.632	-0.701	101.0%	3.97
75%	3	99.1%	3.106	-0.894	100.2%	3.77

We can use either model to generate the Intersection Point and Offset Parameters for a range of Intervention Points and Resource Reduction Factors. The penalty is calculated directly from the duration of the Marching Army and the level of resource:

(Intervention Time – Intersection Time) x Marching Army Resource Level

Some examples of Intersection Points and Offset Parameters are given in Table 2.6. We may have noticed that there is an anomalous pattern in the cost penalties. We might expect that the largest cost penalty occurs with the largest resource reduction, as it does with the schedule penalty, but this is not the case, as illustrated more clearly in Figure 2.13.

Closer examination of this anomalous penalty in Figure 2.14 reveals that the worst case cost penalty varies in the range of approximately 40% to 60% Resource Cap depending on the timing of the Resource Cap Intervention.

For those of us who prefer tabular summaries, Table 2.7 provides a summary of the Cost and Schedule Penalties implied by the Marching Army Technique for a range of Intervention Times and Resource Caps.

(3) Creating a Bactrian Camel Effect (The Disaggregated Perspective)

Sticking with the example that 50% of the resource will be removed prematurely from the project, we could argue that the equivalent of half the outstanding development problems could continue through to the original completion date. However, this still leaves half of the outstanding problems unresolved. Suppose that as the 'continuing half'

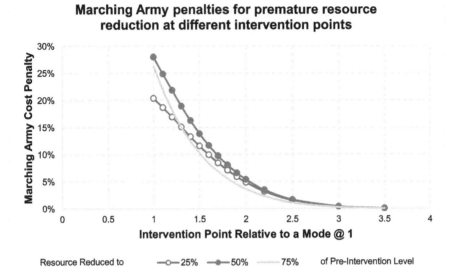

Figure 2.13 Marching Army Penalties for Premature Resource Reduction at Different Intervention Points

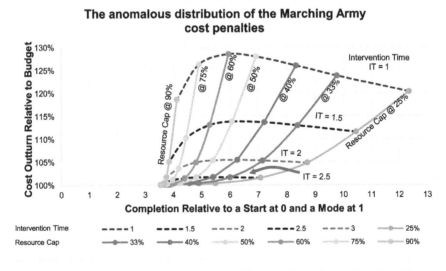

Figure 2.14 The Anomalous Distribution of the Marching Army Cost Penalties

Table 2.7 Marching Army Parameters and Penalties

Intervention Time	25% Resource Cap		50% Resource Cap		75% Resource Cap	
	Cost Outturn	Schedule Outturn	Cost Outturn	Schedule Outturn	Cost Outturn	Schedule Outturn
1	120.35%	12.34	128.01%	6.92	126.29%	4.91
1.5	111.57%	10.45	113.84%	6.07	110.22%	4.45
2	104.86%	8.72	105.38%	5.40	103.63%	4.18
2.5	101.57%	7.07	101.67%	4.80	101.09%	3.97
3	100.39%	5.47	100.41%	4.25	100.26%	3.77

starts to shed its resource as part of the normal development process, we can then retain them to resolve the remaining problems. In essence, we are treating this as if it were a Phased Development Project. (*Yes, we're back to the Camels.*)

Let's assume that the remaining problem resolution follows a Norden-Rayleigh Curve of its own; the question is where is the mode likely to be. We might consider the following four choices:

i. The Mode is the same as the original programme (in this case 1), but is it offset by the point in time that the intervention occurs (in this case 1.5). The mode therefore occurs at period 2.5 and the project end-point is at period 5 (i.e. 3.5 +1.5).

ii. We set the Mode to occur when the maximum resource becomes available, which will be at the end of the original schedule. In this case that would be at 3.5, making the Mode equal to 2 relative to the start point at period 1.5. This implies that the end-point will occur at period 8.5 (i.e. 3.5 x 2 +1.5).

iii. We choose the Mode so that the overall end date is only affected *pro rata* to the resource reduction. In this case there were 2 time periods to go when the resource was reduced by half, implying a doubling of the time to completion. This would imply that a Mode would occur at 1.143 time periods after the intervention (i.e. 4 divided by NRC Truncation Ratio of 3.5), giving a Mode at period 2.643 and an end-point at 5.5.

iv. Pick it at random, it's all guesswork at the end of the day, (*but that's hardly the right spirit, is it?*).

Let's reject the last one as me being a little bit silly and playing 'Devil's Advocate', just testing to see if anyone has dropped off my personal RADAR (*Reader Attention Deficit And Recall*); besides which this option would be completely at odds with the principles of TRACEability (Transparent, Repeatable, Appropriate, Credible and Experientially-based).

For clarity, let's summarise the key parameters and values in Table 2.8 for the first three options in our example. In all cases the NRC End is given by the NRC Start + 3.5 x Mode. The Cost Penalty is the delta cost to the total programme based on the schedule slippage of the deferred work, using the Pro Rata Product Rule.

Table 2.8 Options for Modelling Premature Resource Reduction by Disaggregation (Bactrian Camel)

Value	Original Programme	Deferred Work Option i	Deferred Work Option ii	Deferred Work Option iii
NRC Start	0	1.5	1.5	1.5
NRC Mode	1	2.5	3.5	2.643
NRC End	3.5	5	8.5	5.5
Overall Schedule Penalty	N/A	43%	143%	57%
Overall Cost Penalty	N/A	0%	16%	2.3%

Options i and iii do not 'feel right' suggesting virtually no cost penalty for a dramatic reduction in resource. The cost penalty for Option ii is more in line with the output from our previous Marching Army technique, albeit the schedule slippage is more severe. Figure 2.15 illustrates the Disaggregation Technique for Option ii, with the second hump just being visible in the left-hand graph.

If we were to run this model assuming that the resource cap was applied at time 1 (i.e. the Mode) then the cost penalty at some 43% (see Figure 2.16) would be significantly more than that which we generated using the Marching Army Technique.

Table 2.9 summarises the Cost and Schedule Penalties implied by the Disaggregation Technique for a range of Intervention Times and Resource Caps.

However, the main problem that this technique highlights is that it ignores the interdependence of the development tasks in question. If it is possible to draw this distinction

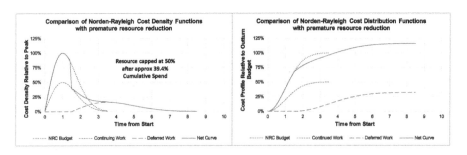

Figure 2.15 Premature Resource Reduction Using a Disaggregation (Bactrian Camel) Technique

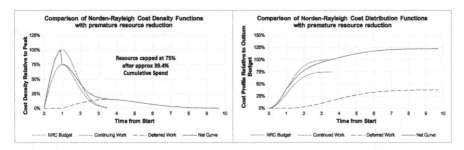

Figure 2.16 Premature Resource Reduction at the Mode Using the Disaggregation Technique

Table 2.9 Summary of Cost and Schedule Penalties Using Disaggregation Technique

Intervention Time Relative to Mode	Cumulative % Completed	90% Resource Cap		75% Resource Cap		50% Resource Cap	
		Cost Outturn	Schedule Outturn	Cost Outturn	Schedule Outturn	Cost Outturn	Schedule Outturn
1 (Mode)	39%	109%	9.75	123%	9.75	145%	9.75
1.5	68%	103%	8.5	108%	8.5	116%	8.5
2	86%	101%	7.25	102%	7.25	103%	7.25
2.5	96%	100%	6	100%	6	100%	6

in outstanding tasks then this may be an appropriate model, especially if we adopt the attitude that towards the project end, the error in any outstanding cost to completion is likely to be very small in the context of what has been achieved previously.

(4) Modal Re-positioning (A Misguided Perspective)

Caveat augur

Don't try this at home ... or at work! The logic is fundamentally flawed.

Why are we even discussing this? Well, it helps to demonstrate that we should always test any theoretical model fully from a logic perspective before trying it with any real data. We could easily mislead or delude ourselves.

The simplest of all possible techniques we could try would be to assume that the reduction to a new level of resource could be interpreted as a new Mode for the revised Norden-Rayleigh Curve for the outstanding work (problem resolution) through to completion. This may sound very appealing to us, as it appears to be very simple.

Let's consider our example again of a resource cap to a 50% level at time = 1.5 (based on the Mode occurring at time = 1). We will assume that we need to double the time to completion from two time periods to four (3.5 minus 1.5, i.e. the difference between the nominal endpoint and the Resource Capping Intervention point). Assuming that the Mode of the outstanding work is pitched at the Intervention Point, then this will give us an overall NRC theoretical duration of 5.6 with a Mode at 1.6, giving us an equivalent offset of 0.1 time units to the left. Table 2.10 and Figure 2.17 illustrates the principle. As we can see, perhaps surprisingly, it returns a very modest penalty of just under 7%.

… However, that's not the worst of it. If we were to increase the intervention point to later in the programme, this penalty reduces even further as shown in Table 2.11.

By flexing the intervention point we can show that there are some fundamental anomalies created using with this technique:

- There is no penalty if we cap the resource at the Mode, which we would probably all agree is nonsensical
- The penalty is independent of the resource cap, i.e. it is the same no matter how much or how little resource is reduced
- A resource reduction just before the end is beneficial – it saves us money. Consequently, this model of progressive intervention should be the model we always use! *Perhaps not*

Table 2.10 Calculation of the Modal Re-Positioning Parameters

Original NRC			Deferred Work		
Nominal Mode @	1				
Nominal Endpoint @	3.5				
Endpoint - Mode	2.5				
Intervention Point @	1.5		1.5	New Mode	
			50%	Resource Cap	
Time to Completion	2		4	Time to Complete with Reduced Resource	
Revised Endpoint			5.5	New Mode + Time to Completion	
Ratio: (End-Start) to (End-Mode)	1.4				
End-Start			5.6		
Start Offset			-0.1	New Start	
			1.6	Time of Mode Relative to Offset Start	

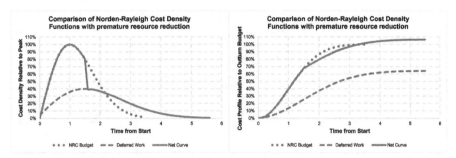

Figure 2.17 Premature Resource Reduction Using the Modal Re-Positioning Technique

For some of us this may be sufficient to reject the technique on the grounds of *reductio ad absurdam*. Consequently, we should completely discard the theory of modelling development resource reduction in this way. It would appear to be fundamentally flawed!

> *Reductio ad absurdum:* Latin phrase use by mathematicians meaning 'reduction to absurdity' – used when demonstrating that an assumption is incorrect if it produces an illogical result (proof by contradiction of an assumption).

(5) Conclusion

There are of course other scenarios that we could be consider, but as estimators the main thing that we need to do is understand the nature of how the project slowdown will be enacted and how we might be able to adapt the principles of Norden-Rayleigh Curves (if at all) to our environment. With that knowledge the informed estimator can choose a logical model that best fits the circumstances. As we demonstrated with the last theoretical model, some models don't stand scrutiny when tested to the limits, or with different parameters.

Please note that no camels have been harmed in the making of this analogy. So, there's no need to get the 'hump' on that account.

2.3 Beta Distribution: A practical alternative to Norden-Rayleigh

The most significant shortcoming of the Norden-Rayleigh Curve is the need to truncate the Rayleigh Distribution CDF at some potentially arbitrary point in time, or in

Table 2.11 Summary of Cost and Schedule Penalties Using the Modal Re-Positioning Technique

Intervention Time Relative to the Mode	25% Resource Cap		50% Resource Cap		75% Resource Cap	
	Cost Outturn	Schedule Outturn	Cost Outturn	Schedule Outturn	Cost Outturn	Schedule Outturn
1	100.00%	11	100.00%	6	100.00%	4.33
1.5	106.59%	9.5	106.59%	5.5	106.59%	4.17
2	102.87%	8	102.87%	5	102.87%	4
2.5	100.20%	6.5	100.20%	4.5	100.20%	3.83
3	99.77%	5	99.77%	4	99.77%	3.67

some cases, the development cycle. This is because the Rayleigh Distribution has an undefined endpoint (*if we describe positive infinity as an unquantifiable right wing concept*).

Caveat augur

There are situations where development programmes have been truncated fairly arbitrarily in the natural development cycle if the number of issues to be resolved are too numerous and complex, and the funding available is insufficient to fulfil the original full development objectives.

In those circumstances, it has been known that a development programme will be stopped and the team takes stock of what has been achieved and what has yet to be demonstrated. These may be known as 'Measure and Declare' projects, and may be sufficient to allow other developments to be started when funding is available

One very flexible group of probability distributions that we reviewed in Volume II Chapter 4 was the family of distributions that we call the Beta Distribution. In addition to their flexibility, Beta Distributions have the added benefit of having fixed start and endpoints.

Let's remind ourselves of the key properties of Beta Distributions, highlighted in Figure 2.18. In this case we may have already seen the similarity between this particular distribution and the characteristic positive skew of the Rayleigh Distribution without the 'flat line' to infinity.

We can use Microsoft Excel's Solver to determine the parameters of the best fit Beta Distribution to the Norden-Rayleigh Curve (i.e. a truncated Rayleigh Distribution). We can see how good (*or indifferent*) the Best Fits are for Truncation Ratios of 2.65 (the Conventional 97% Truncation) and 3.5 (as we have been using in most of our examples),

Probability Density Function for a Beta Distribution

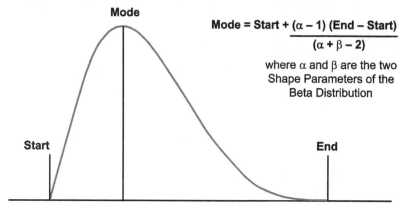

Mode

Mode = Start + (α − 1) (End − Start)

$$\frac{}{(α + β − 2)}$$

where α and β are the two
Shape Parameters of the
Beta Distribution

Start

End

Figure 2.18 Probability Density Function for a Beta Distribution

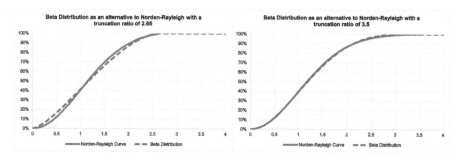

Figure 2.19 How Good is the Best Fit Beta Distribution for a Truncated Norden-Rayleigh Curve?

when we constrain the Sum of Errors to be zero in Figure 2.19. If we are advocates of the conventional Truncation Ratio, then based on this we probably wouldn't use a Beta Distribution as an alternative because the Best Fit is not really a Good Fit. However, if we favour the approach of truncating the Norden-Rayleigh Curve at a higher Ratio (say 3.5) then the two curves are more or less indistinguishable.

In fact because there is some flexibility in where we take the truncation, we can run a series of Solvers to see if there is any pattern between the Beta Distribution parameters and the truncation point. When we set up our Solver and look to minimise the Sum of Squares Error between the Norden-Rayleigh Curve and the Beta Distribution to get the Least Squares Best Fit, we have a choice to make:

a) Do we force the Sum of Errors to be zero?
b) Do we leave the Sum of Errors to be unconstrained?

If we were fitting the Beta Distribution to the original data values then we should be saying 'Option a' to avoid any bias. However, as the Norden-Rayleigh Curve is already an empirical 'best fit curve' with its own intrinsic error, then trying to replicate it with a Beta Distribution is running the risk of over-engineering the fit. True, we would expect that the Best Fit Norden-Rayleigh Curve would pass through the mean of any actual data, and hence the sum of the errors should be zero, but as we have seen in terms of Logarithmic Transformations, this property is not inviolate. However, in deference to the sensitivities of followers of both camps, we will look at it both ways.

If you have been reading some of the other chapters first then you'll be getting familiar with the general approach to our Least Squares Curve Fitting procedure using Microsoft Excel's Solver algorithm. The initial preparation for this is slightly different. As the Outturn value is purely a vertical scale multiplier that applies to both curves, we will only consider fitting the Beta Distribution to the Cumulative Percentage of a Norden-Rayleigh Curve.

1. Decide on the Norden-Rayleigh Curve Truncation Ratio. This is the ratio between the assumed Development end-point relative to the Mode (as discussed in Section 2.2.2) with a given start time of zero, for example 2.65 or 3.5.

2. Using the fact that a Rayleigh Distribution is a special case of the Weibull Distribution with parameters of 2 and Mode $\sqrt{2}$, we can create a Rayleigh Distribution with a Mode of 1 for time values of 0 through to the chosen Truncation Ratio using the Weibull function in Excel: **WEIBULL.DIST(*time*, 2, SQRT(2), TRUE)**. In earlier versions of Microsoft Excel, the function is simply **WEIBULL** with the same parameter structure.

3. We now need to calculate the Rayleigh Distribution Cumulative value at the Truncation Ratio Point, e.g. 2.65 or 3.5, or whatever we decided at Step 1. We use this to divide into the Rayleigh Distribution values created at step 2 in order to uplift the curve to attain 100% at the Truncation Ratio Point. This uplifted Curve is our 'Norden-Rayleigh Curve'.

4. We can create a Beta Distribution using a random pair of alpha and beta parameters but with defined start and end-points of 0 and a value equal to the selected Truncation Ratio Point chosen at Step 1 and enforced at Step 3.

5. We can then calculate the difference or error between the Norden-Rayleigh Curve and the Beta Distribution at various incremental time points.

6. Using Microsoft Excel Solver, we can set the objective of minimising the Sum of Squares Error by changing the two Beta Distribution parameters chosen at random in Step 4.

7. We should aim to add appropriate constraints such as ensuring the Mode of the Beta Distribution occurs at value 1. In this case as the Start Point is zero, the Beta Distribution Mode occurs at $(\alpha - 1) \times$ Truncation Ratio Point$/(\alpha + \beta - 2)$.

8. We should also set the constraint that both the Beta Distribution parameters are greater than 1 to avoid some 'unnatural' results such as twin telegraph poles or

washing line effect (see Volume II Chapter 4). We also have that option to consider of forcing the Sum of Errors to be zero.

9. Finally, we mustn't forget to uncheck the tick box for 'Make Unconstrained Variables Non-Negative'.

Table 2.12 illustrates Steps 1–5 and 7 up to the Mode only but in the model extends through to Time equalling the Norden-Rayleigh Truncation Ratio.

In Table 2.13 we show the results for a range of Norden-Rayleigh Curve (NRC) Truncation Ratios, with and without the Sum of Errors constraint. Figure 2.20 compares the two alternatives. (Note that the Sum of Errors and the Sum of Squares Error are based on time increments of 0.05 from zero through to the Truncation Ratio value.)

Table 2.12 Best Fit Beta Distribution for a Truncated Rayleigh Distribution Using Excel's Solver

	Time	Rayleigh Distribution		Beta Distribution	
	alpha	2		2.143	
	beta	1.414		3.857	
Norden-Rayleigh Truncation Ratio >	Start	0		0	
	End	3.50	>>>	3.50	
	Mode	1		1.00	= Start + End Ratio x (alpha -1)/(alpha + beta - 2)

Time	Rayleigh Distribution CDF	Norden-Rayleigh Uplift		
3.50	99.781%	100.219%		

Time	Rayleigh Distribution	Norden-Rayleigh Curve	Beta Distribution	Error
0	0.000%	0.000%	0.000%	0.000%
0.05	0.125%	0.129%	0.606%	-0.478%
0.1	0.499%	0.514%	1.668%	-1.153%
0.15	1.119%	1.153%	3.002%	-1.849%
0.2	1.980%	2.041%	4.544%	-2.503%
0.25	3.077%	3.171%	6.253%	-3.081%
0.3	4.400%	4.536%	8.102%	-3.566%
0.35	5.941%	6.124%	10.070%	-3.946%
0.4	7.688%	7.925%	12.141%	-4.216%
0.45	9.629%	9.926%	14.300%	-4.374%
0.5	11.750%	12.112%	16.536%	-4.424%
0.55	14.037%	14.469%	18.837%	-4.369%
0.6	16.473%	16.980%	21.196%	-4.216%
0.65	19.043%	19.629%	23.604%	-3.975%
0.7	21.730%	22.398%	26.052%	-3.654%
0.75	24.516%	25.271%	28.534%	-3.264%
0.8	27.385%	28.228%	31.044%	-2.816%
0.85	30.320%	31.253%	33.575%	-2.322%
0.9	33.302%	34.327%	36.122%	-1.794%
0.95	36.317%	37.435%	38.679%	-1.244%
1	39.347%	40.558%	41.241%	-0.683%
Continues through until Time = End Ratio				

Table 2.13 Solver Results for the Best Fit Beta Distribution for a Range of NRC Truncation Ratios

NRC Truncation Ratio	2.5	2.65	3	3.25	3.44	3.5	3.66	4	4.5
With Sum of Errors = 0									
Beta Distribution α	1.313	1.472	1.781	1.976	2.116	2.158	2.288	2.482	2.750
Beta Distribution β	1.469	1.779	2.562	3.197	3.722	3.895	4.372	5.445	7.125
Sum of α and β	2.782	3.252	4.343	5.173	5.838	6.053	6.639	7.927	9.875
Sum of Squares Error	0.0647	0.0374	0.0116	0.0051	0.0029	0.0025	0.0021	0.0033	0.0075
Without Sum of Errors = 0									
Beta Distribution α	1.642	1.714	1.912	2.057	2.164	2.196	2.281	2.444	2.647
Beta Distribution β	1.963	2.179	2.824	3.378	3.839	3.991	4.406	5.332	6.764
Sum of α and β	3.605	3.893	4.735	5.435	6.003	6.187	6.687	7.776	9.411
Sum of Squares Error	0.0209	0.0141	0.0055	0.0029	0.0022	0.0021	0.0021	0.0030	0.0054
Sum of Errors	-82.043%	-65.883%	-38.169%	-23.317%	-13.582%	-10.719%	-3.541%	9.744%	24.966%

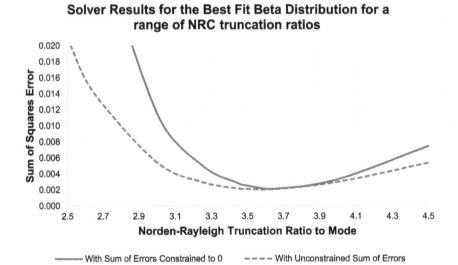

Figure 2.20 Solver Results for the Best Fit Beta Distribution for a Range of NRC truncation Ratios

The best fit curve with the Sum of Errors being constrained to zero occurs around a Norden-Rayleigh Truncation Ratio of 3.66. With an unconstrained Sum of Errors, the ratio is a little lower between 3.5 and 3.66. From Table 2.13 we can see that the sum of the two Beta Distribution parameters for the 'lower Least Square Error' (*if that's not an oxymoron*) is in the region of 6.0 to 6.7 ... *hold that thought for a moment ...*

2.3.1 PERT-Beta Distribution: A viable alternative to Norden-Rayleigh?

Within the Beta family of distributions there is the PERT–Beta group, in which the sum of the two distribution shape parameters is six. These are synonymous with the Cost and Schedule research of the 1950s that gave rise to PERT analysis (Program Evaluation and Review Technique) (Fazar, 1959). Bearing in mind that Norden published his research in 1963, we might speculate that it is quite conceivable that Norden was unaware of the work of Fazar (*it was pre-internet after all*), and hence may not have considered a PERT–Beta Distribution. This line of thought then poses the question 'What if the Solution Development Curve could or should be represented by a PERT–Beta Distribution instead, rather than a Norden–Rayleigh Curve?' What would be the best parameters? Well, running Solver with the additional constraint on the sum of the two parameters being 6 and constraining the Truncation Ratio to 3.5, we get Figure 2.21 in which $\alpha = 2.143$ and $\beta = 3.857$.

> Note: If we prefer to think or recall in fractions, then $\alpha = 2\,^1/_7$ and $\beta = 3\,^6/_7$ with 7 being twice the Norden–Rayleigh Truncation Ratio of 3.5.
>
> If we were to use the 'convention' of a 2.65 ratio, this would give PERT–Beta parameters of $\alpha = 2.51$ and $\beta = 3.49$. A ratio of $2\,^2/_3$ would give PERT–Beta parameters of $\alpha = 2.5$ and $\beta = 3.5$ (*Notice how estimators always gravitate towards rounded numbers?*)

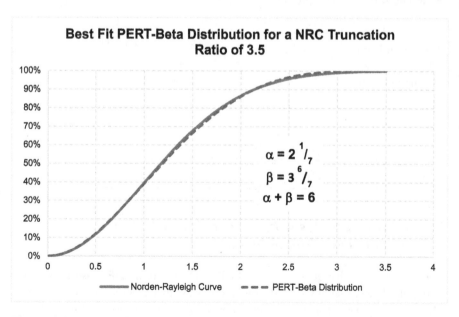

Figure 2.21 Best Fit PERT-Beta Distribution for a NRC Truncation Ratio of 3.5

As we can see the relative difference between the PERT-Beta Distribution and the more general Beta Distribution alternative for a Norden-Rayleigh Curve is insignificant, especially in the context of the likely variation around the curve that we will probably observe when the project is 'actualised'. The beauty of the PERT-Beta analogy with the Norden-Rayleigh is that they are both synonymous with Cost and Schedule analysis.

For the Formula-philes: PERT-Beta equivalent of a Norden-Rayleigh Curve

Consider a PERT-Beta Distribution with parameters α and β, with a start time of A and a completion time of B

For any PERT-Beta Distribution:

$$\alpha + \beta = 6 \qquad (1)$$

From the properties of any Beta Distribution the Mode λ occurs at:

$$\lambda = A + (\alpha - 1)\frac{(B - A)}{(\alpha + \beta - 2)} \qquad (2)$$

Suppose the Start Point A = 0, substituting (1) in (2) and simplifying:

$$\lambda = (\alpha - 1)\frac{B}{4} \qquad (3)$$

Re-arranging (4), the Ratio of the Endpoint to the Mode is given by:

$$\frac{B}{\lambda} = \frac{4}{(\alpha - 1)} \qquad (4)$$

If we take the Norden-Rayleigh Truncation Ratio of 3.5, (4) becomes:

$$3.5 = \frac{4}{(\alpha - 1)} \qquad (5)$$

Solving (5) for α:

$$\alpha = \frac{4}{3.5} + 1 = 2\tfrac{1}{7} \qquad (6)$$

From (1) and (6):

$$\beta = 3\tfrac{6}{7}$$

We can use (4) to derive the equivalent PERT-Beta parameters for any chosen Norden-Rayleigh Curve Truncation Ratio.

2.3.2 Resource profiles with Norden-Rayleigh Curves and Beta Distribution PDFs

Figure 2.22 illustrates that our Best Fit PERT-Beta Distribution PDF is also a reasonable fit to the Norden-Rayleigh Curve Resource Profile, with a Truncation Ratio of 3.5 (*to avoid that nasty step-down. For those of us who favour the step-down, we may need to consider how we deal with the inevitable cost creep that we often observe, potentially from external purchases*).

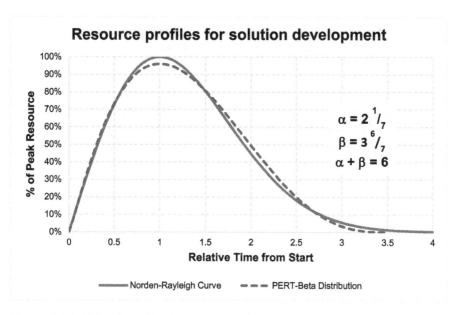

Figure 2.22 PERT-Beta Distribution PDF cf. Norden-Rayleigh Resource Profile

We could also use Microsoft Excel's Solver to model a Beta Distribution PDF (general or PERT) to the cost expenditure per period. However, we may find that the data is more erratic whereas the Cumulative Distribution Function approach does dampen random variations, often caused by external expenditure or surge in overtime etc.

2.4 Triangular Distribution: Another alternative to Norden-Rayleigh

If we look at a Norden-Rayleigh Curve we may conclude that it is little more than a weather-worn Triangular Distribution ... essentially triangular in shape but with rounded off corners and sides. In fact, just as we fitted a Beta Distribution to represent a NRC, we could fit a Triangular Distribution ... only not as well, as illustrated in Figure 2.23.

The appeal of the Triangular Distribution is its simplicity; we only need three points to define its shape so for a quick representation of resource requirements, we may find it quite useful. The downside of using it is the fact that Microsoft Excel is not 'Triangular Friendly' in that there is no pre-defined function for it and we have to generate the calculations long-hand (see Volume II Chapter 4).

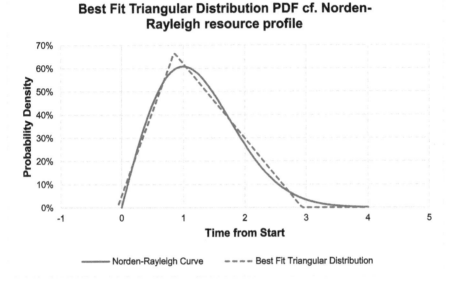

Figure 2.23 Best Fit Triangular Distribution PDF cf. Norden-Rayleigh Resource Profile

Most of us will probably have also spotted the key difference between a 'pure' Norden-Rayleigh Curve and the Best Fit Triangular Distribution (*don't worry if you didn't*) ... the latter is offset to the left in terms of the three parameters that define a Triangular Distribution (Minimum, Mode and Maximum) as summarised in Table 2.14:

These may seem like some fundamental differences in the two sets of parameters but we should ask ourselves how precise we think the original research and analysis really was that spawned the Norden-Rayleigh Curve as a truncation of the Rayleigh Distribution? (*This is not meant to be taken as a criticism of their valuable contribution.*) The question we should be posing is not one of precision but one of appropriate level of accuracy. If we compare the two Cumulative Distribution Functions, we will see that they are virtually indistinguishable (Figure 2.24).

2.5 Truncated Weibull Distributions and their Beta equivalents

2.5.1 Truncated Weibull Distributions for solution development

As we have said already, the Rayleigh Distribution is just a special case of the more generalised Weibull Distribution. Let's turn our thinking now to whether that can be utilised for profiling Solution Development costs in a wider context.

Table 2.14 Triangular Distribution Approximation to a Norden–Rayleigh Curve

Distribution	Norden-Rayleigh	Triangular
Minimum	0	-0.07
Mode	1	0.856
Maximum	3.5	2.93
Range (Max-Min)	3.5	3
Mode-Min	1	0.927
(Max-Min) / (Mode-Min)	3.5	3.238

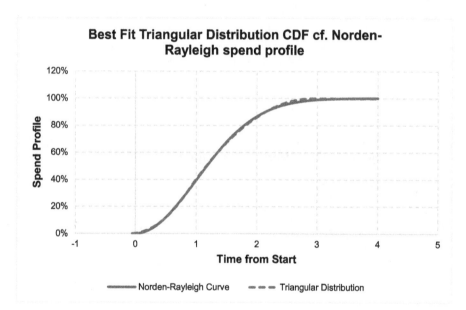

Figure 2.24 Best Fit Triangular Distribution CDF cf. Norden–Rayleigh Spend Profile

Let's consider two related development programmes running concurrently, each of which 'obeys' the Norden–Rayleigh criteria proposed by Norden (1963) in Section 2.1. Suppose further that the two developments are slightly out of phase with each other, but not so much as to create one of our Bactrian Camel effects from Section 2.2.1. In essence, we just have a long-backed Dromedary Camel effect.

Suppose we have two such developments that follow the Norden Rayleigh Curve (NRC) patterns of behaviour:

- One development project (NRC 1) commences at time 0 and has a peak resource at time 1.25, finishing at time 4.375 (i.e. 3.5 x 1.25). Let's suppose that this development generates 70% of the total development cost
- The second development project (NRC 2) commences at time 0.75, peaks at time 2.25, has a nominal completion at time 6 (i.e. $3.5\times(2.25-0.75)+0.75$), and of course accounts for the remaining 30% of the work.

Using Microsoft Excel's Solver, we can generate a Weibull Distribution that closely matches the sum of the two NRCs as shown in Figure 2.25. In this case the Weibull Distribution has parameters $\alpha = 2.014$ and $\beta = 2.1029$.

However, this could easily be overlooked as being a more general Weibull Distribution, as it is so close to a true Norden–Rayleigh Curve. The first parameter of 2.014 being a bit of a giveaway and could be taken as 2 as would be the case for the true NRC without any material effect on the curve error. In fact, the Best Fit NRC for the combined development would commence at time 0 and have a β Parameter of 2.1, giving a mode of 1.485 and an implied finishing time at 5.2. (For a Norden–Rayleigh Curve, the Weibull parameters are $\alpha = 2$ and $\beta = \text{Mode} \times \sqrt{2}$.) In the context of estimating, the difference is insignificant, implying perhaps that the Norden–Rayleigh Curve 'rules' of a single set of integrated development objectives are not sacrosanct, and that there is some

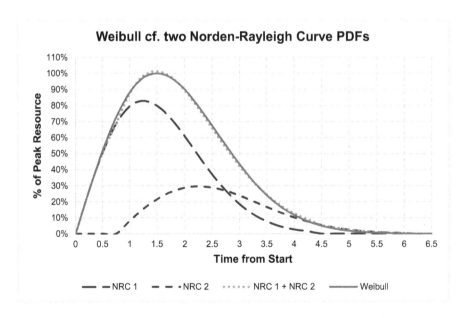

Figure 2.25 Best Fit Weibull Distribution for the Sum of Two Norden–Rayleigh Curves (1)

pragmatic flexibility there! Under certain conditions, they can be relaxed, indicating why the basic shape of a Norden-Rayleigh Curve may seem to fit larger development projects with evolving objectives such as some major defense platforms, as we saw with the F/A-18 and F/A-22 in Section 2.1.2.

Now let's consider the same two NRCs but with a greater overlap due to the second project slipping to the right by half a time period. (*In terms of the inherent 'camelling effect' we still have a basic single hump Dromedary Camel's profile and not a double hump Bactrian Camel*):

- The first development project (NRC 1) still commences at time 0 and has a peak resource at time 1.25, finishing nominally at time 4.375. Let's suppose that this development generates 70% of the total development cost.
- The second development project (NRC 2) now commences at time 1.25, peaks at time 2.75, with a nominal finishing at time 6.5. Again, this accounts for the remaining 30% of the development work.

These can be approximated by a general Weibull Distribution with parameters $\alpha = 1.869$ and $\beta = 2.283$. Clearly not a Norden-Rayleigh Curve (see Figure 2.26).

In reality we will always get variances between our actuals in comparison to any empirical model. This will be the case here, and any such variances could easily hide or disguise the better theoretical model. It may not matter in the scheme of things but this illustrates that models should be viewed as an aid to the estimator and not as a mathematical or intellectual straitjacket.

2.5.2 General Beta Distributions for solution development

Just as we demonstrated in Section 2.4 where we could substitute a particular PERT-Beta Distribution for a Norden-Rayleigh Curve, we can just as easily substitute a truncated Weibull distribution with a general Beta Distribution. The benefit is that we don't have to bother with all that truncation malarkey as the Beta Distribution has fixed start and endpoints.

If we were to re-run our Excel Solver models for the two scenarios in the preceding section but substitute a Beta Distribution for the Weibull Distribution as an approximation for the sum of two NRCs, then we will get the results in Figures 2.27 and 2.28. Note: we may need to add a constraint that prevents the Model choosing Parameter values equalling 1, or forcing the model to take a value of 0 at the start time as Excel's **BETA.DIST** function can return errors in some circumstances.

The Beta Distribution parameters that these generate are as follows:

- Case 1: $\alpha = 2.408$ $\beta = 4.521$ $\alpha + \beta = 6.930$
- Case 2: $\alpha = 1.997$ $\beta = 3.493$ $\alpha + \beta = 5.490$

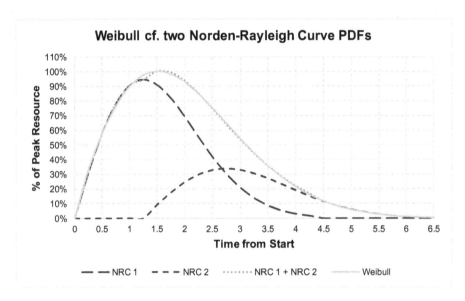

Figure 2.26 Best Fit Weibull Distribution for the Sum of Two Norden-Rayleigh Curves (2)

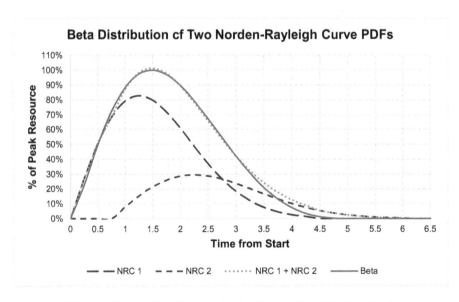

Figure 2.27 Best Fit Beta Distribution for the Sum of Two Norden-Rayleigh Curves (1)

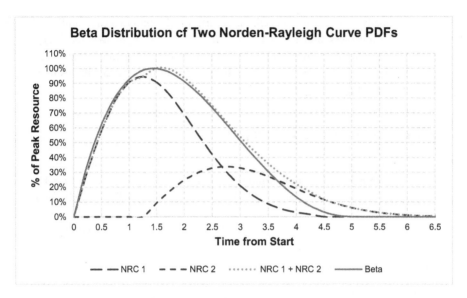

Figure 2.28 Best Fit Beta Distribution for the Sum of Two Norden-Rayleigh Curves (2)

Note: Unlike the Beta Distribution approximation for a simple Norden-Rayleigh Curve, the sum of the two parameters are distinctly not 6. Hence, in these cases a PERT-Beta Distribution is unlikely to be an appropriate approximation model.

2.6 Estimates to Completion with Norden-Rayleigh Curves

Where we have an ongoing research and development project, it is only to be expected that we will want to know what the likely outturn will be, i.e. when will we finish, and at what cost?

Let's assume that our project follows the criteria expected of a Norden-Rayleigh Curve, and that we have been tracking cost and progress with an Earned Value Management System. How soon could we realistically create an Estimate or Forecast At Complete (EAC or FAC)?

In the view of Christensen and Rees (2002), earned value data should be sufficient to fit a Norden-Rayleigh Curve after a development project has achieved a level of 20% completion or more. Although this 20% point is not demonstrated empirically, the authors believe that EACs are sufficiently stable after this. However, under the premise suggested by Norden, the problems and issues to be resolved as an inherent part of the development process, are still being identified at this point faster than they can be resolved, hence the need to increase the resource. Once we have achieved the peak resource (nominally at the 40% achievement point), we may find that the stability of the EAC/FAC only improves after this. (*After this, it's all downhill, figuratively speaking.*) For the time being, just hold this thought, as we will return to it at the end of Section 2.6.3.

We have a number of options available to us in terms of how we tackle this requirement, as summarised in Table 2.15.

2.6.1 Guess and Iterate Technique

This is the simplest technique, but not really the most reliable, so we won't spend much time on it. We can use it with either a pure Norden-Rayleigh Curve, or its PERT-Beta Distribution lookalike. In terms of our TRACEability paradigm (Transparent, Repeatable, Appropriate, Credible and Experientially-based), it fails on Repeatability. True, we can repeat the technique but we cannot say that another estimator would necessarily come up with the same (or even similar) results.

We'll demonstrate the 'hit and miss nature' of this technique with an example using both a Norden-Rayleigh Curve and a PERT-Beta lookalike. Figure 2.29 illustrates our progress against a budget profiled using a Norden-Rayleigh Curve. Clearly we are not following the budget, having started late and now apparently spending at a higher rate than planned.

In this case we have varied the Start and Endpoints, and the Cost Outturn values until we have got what appears to be a good visual fit. The values we have settled on are:

Start = 3
End = 24
Outturn = € 6,123 k

Table 2.15 Options for Creating EACs for Norden-Rayleigh Curves

Basic Technique	Truncated Rayleigh Distribution	Beta Distribution Lookalike	Comment
1. Guess and Iterate and judge the goodness of fit by the "rack of eye"	✓	✓	This could also be called the "Hit and Miss" Technique
2. Curve Fitting and Extrapolation with Microsoft Excel Solver	✓	✓	Using the principle of Least Squares Error
3. Linear Transformation, Linear Regression	✓	✗	Using the principle of Least Squares backed up by measures of statistical significance
4. Curve Fitting and Extrapolation exploiting Weibull's Double Log Linearisation	✓	✗	Similar to Option 2 but with an added constraint on the Least Squares algorithm

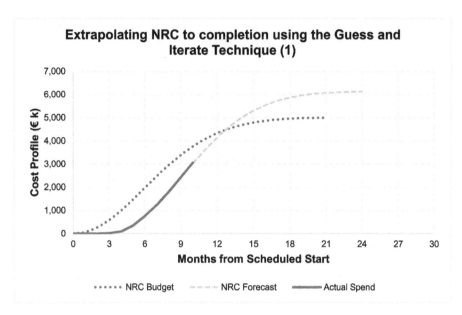

Figure 2.29 Extrapolating NRC to Completion Using the Guess and Iterate Technique (1)

Note that we could equally have chosen the Mode instead of the Endpoint as one of our parameters ... or even the Mode instead of our Start point.

However, an alternative 'Guess' from a different iteration shown in Figure 2.30 looks just as convincing, and we're not just talking about a small sensitivity here. The values here are:

$$Start = 3$$
$$End = 30$$
$$Outturn = € 8,250 k$$

Whilst fitting the cumulative curve to the cumulative actuals to date has the benefit of smoothing out any random variations between individual time periods, it does rather suggest that we can fit a whole range of Norden-Rayleigh Curves through the data that looks good enough to the naked eye. What instead if we were to look at the equivalent spend per month data instead? Figure 2.31 does just that for us for both the parameter 'Guess' iterations above. The left-hand graph, which corresponds to our first guess, appears to suggest that the mode occurs to the right of where our guesses at the Start and Endpoints have implied (using the 3.5 Ratio rule for Mode relative to the End). However, the right-hand graph has some bigger variances on the ramp-up than the left-hand graph. (*This could take some time to get a better result if we want to continue iterating!*)

However, in fitting the model to the monthly spend patterns we must not let ourselves be drawn into using the PDF version of the NRC ...

Caveat augur

If we choose to use the 'Guess and Iterate' (or the Microsoft Excel Solver Technique, for that matter) on the Monthly Spend profile rather than the Cumulative Spend profile, we should avoid fitting the data to a model based on the Probability Density Function of the Weibull Distribution. The PDF gives us the 'spot value' of cost, or 'burn rate' at a point in time, not the cost spend between points in time.

The way around this is to disaggregate the Cumulative Curve taking the difference between cumulative values of each pair of successive time periods

What if we were to try the same thing using the PERT-Beta lookalike? Would it be still as volatile? In Figure 2.32 we have simply used the same parameter 'Guesses' as we did in the two iterations above using the 'pure' Norden-Rayleigh Curve. Whilst the cumulative curve in the top-left appears to be a very good fit to the actual data, the equivalent Monthly Spend below it suggests that we have been too optimistic with the Endpoint and the Outturn Value. In contrast the right-hand pair of graphs suggest that the Endpoint and Outturn values are too pessimistic.

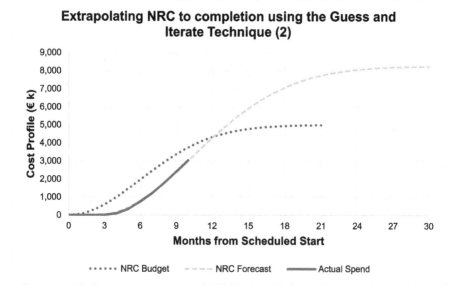

Figure 2.30 Extrapolating NRC to Completion Using the Guess and Iterate Technique (2)

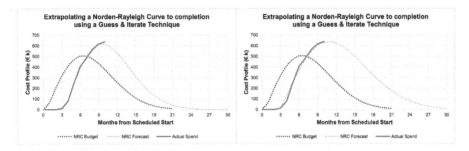

Figure 2.31 Extrapolating NRC to Completion Using the Guess and Iterate Technique (3)

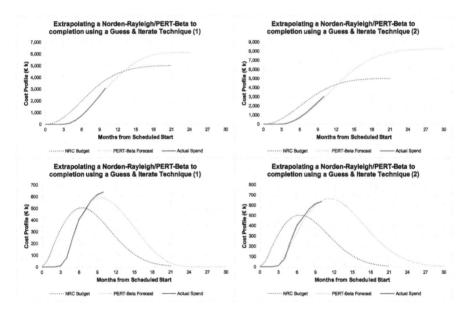

Figure 2.32 Extrapolating a PERT-Beta Lookalike to Completion Using a Guess and Iterate Technique (1)

If instead we were to try to 'Guess and Iterate' values between these two we might conclude that the results in Figure 2.33 are better, and importantly that the range of outturn values for both cost and schedule have narrowed:

Parameter	Left	Right
Start	3	3
End	25.5	27
Outturn	€ 7,000 k	€ 7,100 k

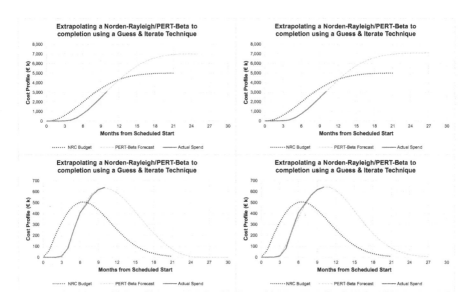

Figure 2.33 Extrapolating a PERT-Beta Lookalike to Completion Using a Guess and Iterate Technique (2)

However, there is both a risk and an opportunity in the psychology of using this technique (with either an NRC or PERT-Beta.) There is a risk that we tend to iterate with fairly round numbers and miss the better fit that we might get by using more precise values for the three parameters. Paradoxically, therein lies the opportunity. We avoid going for the perfect best fit answer because we know that life doesn't work that way; performance will change so we should not delude ourselves into sticking unswervingly to the current trend as the absolute trends do and will change, and consequently, so will the EAC. What we should be looking for is some stability over time.

2.6.2 Norden-Rayleigh Curve fitting with Microsoft Excel Solver

This technique will give us that more precise, albeit not necessarily more accurate result. The main benefit over 'Guess and Iterate' is its speed and repeatability. Appropriateness and Explicitness comes in the definition of any constraints we choose to impose, thus meeting our TRACEability paradigm.

In Table 2.16 we show a Model Set-up that allows Microsoft Excel's Solver to vary the Start, Endpoint and Cost Outturn by minimising the Sum of Squares Errors

Table 2.16 Solver Model Set-Up for Norden-Rayleigh Curve Forecast

Parameter	Rayleigh Distribution Budget	Rayleigh Distribution Forecast
NRC Factor	3.5	
Start	0	0.00
Mode	6	6.00
End	21	21.00
√2	1.414	1.414
alpha	2	2
beta	8.485	8.485
Total Budget	€ 5,000 k	€ 5,000 k

Nominal End Month	Rayleigh Distribution CDF	Norden-Rayleigh Uplift
21.00	99.781%	100.219%

Month	Rayleigh Distribution	NRC Budget	Actual Spend	NRC Forecast	Model Error
0	0.000%	€ 0 k	€ 0 k	€ 0 k	€ 0 k
1	1.379%	€ 69 k	€ 0 k	€ 69 k	-€ 69 k
2	5.404%	€ 271 k	€ 0 k	€ 271 k	-€ 271 k
3	11.750%	€ 589 k	€ 12 k	€ 589 k	-€ 577 k
4	19.926%	€ 998 k	€ 91 k	€ 998 k	-€ 907 k
5	29.335%	€ 1,470 k	€ 346 k	€ 1,470 k	-€ 1,124 k
6	39.347%	€ 1,972 k	€ 752 k	€ 1,972 k	-€ 1,220 k
7	49.366%	€ 2,474 k	€ 1,234 k	€ 2,474 k	-€ 1,240 k
8	58.889%	€ 2,951 k	€ 1,799 k	€ 2,951 k	-€ 1,152 k
9	67.535%	€ 3,384 k	€ 2,415 k	€ 3,384 k	-€ 969 k
10	75.065%	€ 3,761 k	€ 3,053 k	€ 3,761 k	-€ 708 k

			Sum of Errors	-€ 8,237 k
			Sum of Squares Error	8290204

between the Actual Cumulative Spend and the Cumulative NRC Forecast Model. We have some options for the constraints that we choose to impose. Generally we would expect to set the constraint that the Sum of the Errors should be zero in line with usual practice for Least Squares Error, but there are occasions where we will get a 'better fit' if we relax that constraint, especially if we feel that there is already inherent bias in the model towards the Start point. We can also exercise our judgement to limit the range of possible values for Solver to consider in terms of the parameters; for instance, in this case we might want to specify that the Start point must be no less than 2 … and again we shouldn't forget to untick the box marked 'Make Unconstrained Values Non-negative'. Here, we have taken the starting parameter guesses to be the budget parameters; the results shown in Table 2.17 and Figure 2.34 relate to those in which the Sum of Errors is zero.

Table 2.17 Solver Model Results for Norden–Rayleigh Curve Forecast

Parameter	Rayleigh Distribution Budget				Rayleigh Distribution Forecast
NRC Factor	3.5				
Start	0				2.86
Mode	6				9.40
End	21				25.74
√2	1.414				1.414
alpha	2				2
beta	8.485				9.242
Total Budget	€ 5,000 k		Outturn		€ 6,777 k

Nominal End Month	Rayleigh Distribution CDF	Norden–Rayleigh Uplift
21.00	99.781%	100.219%

Month	Rayleigh Distribution	NRC Budget	Actual Spend	NRC Forecast	Model Error
0	0.000%	€ 0 k	€ 0 k	#N/A	€ 0 k
1	1.379%	€ 69 k	€ 0 k	#N/A	€ 0 k
2	5.404%	€ 271 k	€ 0 k	#N/A	€ 0 k
3	11.750%	€ 589 k	€ 12 k	€ 1 k	€ 11 k
4	19.926%	€ 998 k	€ 91 k	€ 102 k	-€ 11 k
5	29.335%	€ 1,470 k	€ 346 k	€ 353 k	-€ 7 k
6	39.347%	€ 1,972 k	€ 752 k	€ 738 k	€ 14 k
7	49.366%	€ 2,474 k	€ 1,234 k	€ 1,232 k	€ 2 k
8	58.889%	€ 2,951 k	€ 1,799 k	€ 1,804 k	-€ 5 k
9	67.535%	€ 3,384 k	€ 2,415 k	€ 2,421 k	-€ 6 k
10	75.065%	€ 3,761 k	€ 3,053 k	€ 3,050 k	€ 3 k

			Sum of Errors		€ 0 k
			Sum of Squares Error		535

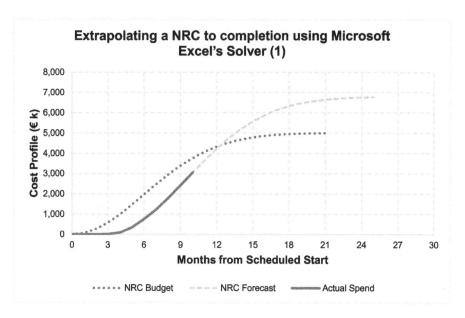

Figure 2.34 Extrapolating a NRC to Completion Using Microsoft Excel's Solver (1)

This gives us an Outturn value of € 6,777 k based on a Start point at month 2.86 and an Endpoint at month 25.74 (with a Mode at month 9.4).

We may recall from the previous technique that whilst we had what appeared to be a good fit to the cumulative profile, when we looked at it from a monthly perspective, it was not such a good result. Solver will overcome this in general, if it can, as illustrated in Figure 2.35.

Let's see what happens if we use Solver with the NRC PERT-Beta lookalike. The model set-up is similar to the above but uses a **BETA.DIST** function instead with fixed parameters $\alpha = 2\,^1/_7$ and $\beta = 3\,^6/_7$. The Solver variable parameters are as before, and our Solver results are shown in Figures 2.36 and 2.37, and are based on the following parameters:

Start = 2.88

End = 27.51

Outturn = € 7,697 k

This result also appears to be an excellent fit to the data from both a cumulative and monthly burn rate perspective. However, it is fundamentally different to the forecast to completion created using the 'pure' Norden-Rayleigh Curve Solver model. Not only is it different, but it is substantially different! (*Put that bottom lip away and stop sulking.*) Let's look at why and how we can use this to our benefit rather than detriment.

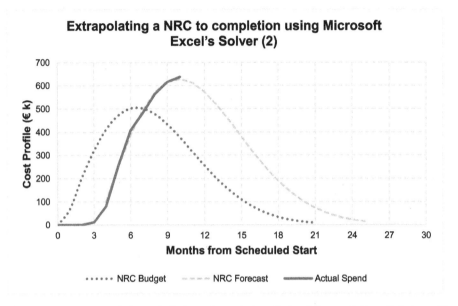

Figure 2.35 Extrapolating a NRC to Completion Using Microsoft Excel's Solver (2)

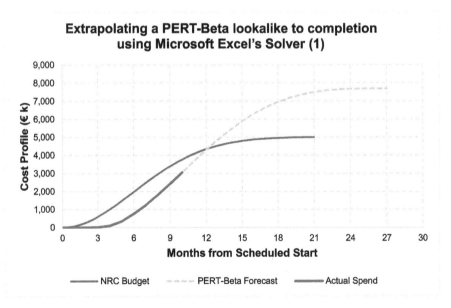

Figure 2.36 Extrapolating a PERT-Beta Lookalike to Completion Using Microsoft Excel's Solver (1)

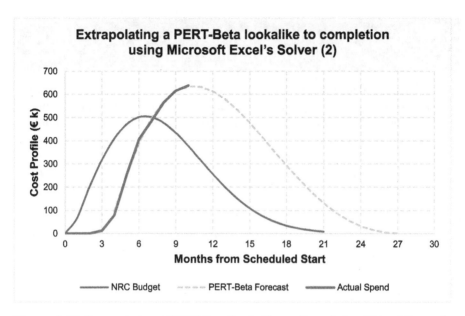

Figure 2.37 Extrapolating a PERT-Beta Lookalike to Completion Using Microsoft Excel's Solver (2)

If we recall from Figure 2.21 in Section 2.3.1, we saw the PERT-Beta Distribution was an excellent substitute for a Norden-Rayleigh Curve. However, in Figure 2.22 in Section 2.3.2 we showed that whilst the mode occurs in the same place there is a difference in the Probability Density Functions of the two distributions. The PERT-Beta has a slightly smaller Modal height, which is compensated for elsewhere. As our actuals are only available until around the Mode this will create a measurable difference between the two models. Look on the bright side … we now have a range estimate; the Norden-Rayleigh Model gives us a more optimistic prediction in this position than the PERT-Beta lookalike.

Once we get past the mode, things should begin to stabilise more and the gap will begin to narrow (assuming that this is a development project that follows the 'rule' of a single, bounded development as described in Section 2.1.4 to a reasonable extent).

2.6.3 Linear transformation and regression

Now from the look on some of our faces, we may be wondering how on Earth we can transform something like a Norden-Rayleigh Curve into straight line.

For the Formula-philes: Simple linear representation of a Norden-Rayleigh Curve

Consider a Norden-Rayleigh Curve generated from a Rayleigh Distribution with a Start point at time 0, a Mode at time λ and a nominal completion time at time τ, giving an uplift factor of k relative to the corresponding Rayleigh Distribution, in order to get an Outturn Cost of C_T

Cumulative Norden-Rayleigh Curve $N_\lambda(t)$ at time t:

$$N_\lambda(t) = kC_T\left(1 - e^{-\frac{1}{2}\left(\frac{t}{\lambda}\right)^2}\right) \qquad (1)$$

Re-arranging (1):

$$kC_T - N_\lambda(t) = kC_T e^{-\frac{1}{2}\left(\frac{t}{\lambda}\right)^2} \qquad (2)$$

Simplifying (2) to give the percentage of work outstanding:

$$1 - \frac{N_\lambda(t)}{kC_T} = e^{-\frac{1}{2}\left(\frac{t}{\lambda}\right)^2} \qquad (3)$$

Taking the Natural Log of (3), substituting Ln(e) = 1 and re-arranging:

$$\ln\left(1 - \frac{N_\lambda(t)}{kC_T}\right) = -\frac{1}{2}\left(\frac{t}{\lambda}\right)^2 \qquad (4)$$

Substituting s for t^2 as the square of time:

$$\ln\left(1 - \frac{N_\lambda(t)}{kC_T}\right) = -\frac{1}{2\lambda^2}s \qquad (5)$$

Expanding the Natural Log expression and re-arranging:

$$\ln\left(kC_T - N_\lambda(t)\right) = -\frac{1}{2\lambda^2}s + \ln\left(kC_T\right)$$

The product of the NRC Uplift Factor and its Outturn value (kC_T) gives us the equivalent Rayleigh Distribution Estimate At Infinity (or RD-EAI) and the Rayleigh Distribution Estimate To Infinity (or RD-ETI) is the difference between the RD-EAI and the Actuals to Date ($kC_T - N_\lambda(t)$).

Therefore, we can say that the Natural Log of the Rayleigh Distribution Estimate To Infinity is a linear function of the Square of Time Elapsed and the Rayleigh Distribution Estimate At Infinity.

The problem seems to be that we need to know the outturn value in order to perform the transformation … in other words, if we knew the answer then we would be able to solve the problem … which all seems a bit of a pointless circular argument because if we knew the answer then we wouldn't have a problem. *It isn't just me, is it?*

However, before we throw our arms in the air in despair, let's take a step back; all is not lost. Microsoft Excel's Solver can come to our rescue again! In Table 2.18 we show a possible model setup.

1. The left-hand three columns allow us to profile the budget of € 5,000 k as a Norden-Rayleigh Curve commencing at month 0 with a Mode at month 6, giving a nominal endpoint at month 21. In Microsoft Excel this is based on the Rayleigh Distribution as a special case of the Weibull Distribution with parameters of 2 and Mode √2 multiplied by an uplift factor of 100.219%, derived from the Truncation Ratio of 3.5

2. The fourth column shows the actuals to date

3. The header section of the next two columns depict the Solver starting parameters for our Solver model to determine the Best Fit Norden-Rayleigh Curve through the actuals. Here we have taken the budget values for the Start point, Mode and Outturn, although we could have picked any non-negative values at random

4. The NRC Outturn multiplied by the Uplift Factor gives us our theoretical Rayleigh Distribution Estimate At Infinity (RD-EAI), shown as € 5,011 k in the example

Table 2.18 Solver Model Setup for Linear Transformation of a Norden-Rayleigh Curve

Parameter	Rayleigh Distribution			Rayleigh Distribution	Time from Start						
NRC Factor	3.5										
Start	0			0.000	0					Time from Start	
Mode	6			6.000	6.000	>>>>>	>>>>>	>>>>>	>>>>>	Mode	10.331
End	21			21.000	18.160						
√2	1.414			1.414	1.414						
alpha	2			2	2						
beta	8.485			8.485	8.485						
Total Budget	€ 5,000 k		Outturn >	€ 5,000 k						Regression	

Nominal End Month	Rayleigh Distribution CDF	Norden-Rayleigh Uplift		Outturn x NRC Uplift Factor						No of Obs	11
										Slope	-0.005
										Intercept	8.8357
21.00	99.781%	100.219%	RD-EAI >	€ 5,011 k	>>>>>	>>>>>	>>>>>	>>>>>	>>>>>	RD-EAI	€ 6,876 k

Month	Rayleigh Distribution	NRC Budget	Actual Spend	Time from Start (Month - Start)	Best Fit Offset NRC	Error (Actual - Best Fit)	Time from Start Squared (s)	RD-ETI	Natural Log Ln(RD-ETI)	Best Fit Ln(RD-ETI)	Error (Best Fit - Ln(RD-ETI))
0	0.000%	€ 0 k	€ 0 k	0.000	€ 0 k	€ 0 k	0.00	5,000	8.5172	8.8357	-0.3185
1	1.379%	€ 69 k	€ 0 k	1.000	€ 69 k	-€ 69 k	1.00	7,001	8.8538	8.8426	0.0112
2	5.404%	€ 271 k	€ 0 k	2.000	€ 271 k	-€ 271 k	4.00	7,001	8.8538	8.8091	0.0447
3	11.750%	€ 589 k	€ 12 k	3.000	€ 589 k	-€ 577 k	9.00	6,989	8.8521	8.7533	0.0988
4	19.926%	€ 998 k	€ 91 k	4.000	€ 998 k	-€ 907 k	16.00	6,910	8.8407	8.6751	0.1656
5	29.335%	€ 1,470 k	€ 346 k	5.000	€ 1,470 k	-€ 1,124 k	25.00	6,655	8.8031	8.5746	0.2285
6	39.347%	€ 1,972 k	€ 752 k	6.000	€ 1,972 k	-€ 1,220 k	36.00	6,249	8.7402	8.4517	0.2884
7	49.366%	€ 2,474 k	€ 1,234 k	7.000	€ 2,474 k	-€ 1,240 k	49.00	5,767	8.6599	8.3065	0.3534
8	58.889%	€ 2,951 k	€ 1,799 k	8.000	€ 2,951 k	-€ 1,152 k	64.00	5,202	8.5568	8.1390	0.4178
9	67.535%	€ 3,384 k	€ 2,415 k	9.000	€ 3,384 k	-€ 969 k	81.00	4,586	8.4307	7.9491	0.4816
10	75.065%	€ 3,761 k	€ 3,053 k	10.000	€ 3,761 k	-€ 708 k	100.00	3,948	8.2809	7.7369	0.5440
11	81.373%	€ 4,078 k		11.000	€ 4,078 k						
12	86.466%	€ 4,333 k		12.000	€ 4,333 k					Total Error	2.3155
13	90.437%	€ 4,532 k		13.000	€ 4,532 k					SSE	1.103544
14	93.427%	€ 4,682 k		14.000	€ 4,682 k					SSE x 10⁶	1103544

5. We can calculate the new variable 's' as the square of the time from commencement, i.e. the actual month number minus the actual start parameter (not necessarily zero once Solver begins its iterations)

6. We can determine the error between Actuals and the current 'Best Fit' Model allowing errors to be generated where the month is less than the Starting month parameter. (Errors will be generated when the 'Time from Start' value is less than zero.) We will exploit this in Step 8

7. The Rayleigh Distribution Estimate To Infinity column (RD-ETI) is calculated by taking the RD-EAI minus the Actuals to date. We can then take the Natural Log of this in the next column

8. Above these we can now calculate the number of points we should be using in the Regression step; this is simply the number of points which do not generate an error in Step 5. We do this by performing a simple count in Excel using the **COUNT(*range*)** function as it will ignore errors generated at Step 6

9. *Now this is the part where the Excel bit might get a bit stretching.* We can now calculate the Regression Slope and Intercept using a combination of Excel Function **SLOPE** in conjunction with **OFFSET**, and also **INTERCEPT** in conjunction with **OFFSET**

- **SLOPE(*ETI-range, s-range*)** and **INTERCEPT(*ETI-range, s-range*)**
- ***ETI-range*** = **OFFSET(*Blank-ETI,-RegPoints,0,RegPoints,*1)**
- ***s-range*** = **OFFSET(*Blank-s,-RegPoints,0,RegPoints,*1)**
- ***Blank-ETI*** is the first blank cell immediately under the data range in column Natural Log Ln(RD-ETI)
- ***Blank-s*** is the first blank cell immediately under the data range in column Time from Start Squared (s)
- ***RegPoints*** is the number of valid regression points calculated in Step 8

Using the **OFFSET** function here allows us to count back the number of non-negative data points from Step 8 and creates an array or range one column wide and includes the number of rows defined by the number of Regression Points

10. Using the calculated Slope of the regression line between time square variable s and Natural Log of RD-ETI, we can calculate the Offset Mode of the implied NRC as the reciprocal of $(-2 \times \text{Slope})$. The cell RD-EAI can be calculated in Excel as **EXP(*Intercept*)** to reverse the transformation.

11. The penultimate column is the Line of Best Fit using the calculated Intercept and Slope applied to the Squared Time variable s, i.e. Intercept plus Slope multiplied by s

12. The final column expresses the difference between the 'actual' and Best Fit Ln(RD-ETI) columns, i.e. the error terms

13. We can then calculate the Sum of these error values (which should always be zero as we have forced a regression through the data ... if they are not, we need to check

our model out for mistakes). We can determine the Sum of Squares Errors (SSE) using **SUMSQ** in conjunction with **OFFSET**

- **SUMSQ**(*Error-range*)
- *Error-range* = **OFFSET**(*Blank-Error,-RegPoints,0,RegPoints,1*)
- *Blank-Error* is the first blank cell immediately under the data range in the end column
- *RegPoints* is the number of valid regression points calculated in Step 8

14. Now for a little trick to make Excel Solver work that little bit harder for us. SSE is our Solver objective to minimise, but it will only try to minimise to a specified number of decimal places. As we are dealing with relatively small numbers with the Natural Log of the error values, we can force a more stringent convergence requirement in the Solver options … or we can take the lazy option and simply multiply the SSE by a fixed large number, like a million. (*'Efficiency is intelligent laziness'* as they say)

15. Finally, we can set up our Solver model using the inflated SSE as the objective to minimise by varying the Start, Mode and Outturn parameters subject to the following constraints:

- The two calculations in the header section for the theoretical Rayleigh Distribution Estimate At Infinity, RD-EAI, are the same (i.e. fifth column and the last column, highlighted by chevrons)
- The calculated Mode based on the Regression Slope equals the Solver input variable (sixth column and last column of the header section, highlighted by chevrons)
- The Outturn is greater than or equal to the actuals to date
- The Project Endpoint is greater than current month for which we have actual data

Ensure that the 'Make Unconstrained Variables Non-negative' option is not selected

We can now run Solver … only to find that it fails to converge in this instance. (*Now, now, there's no need to use language like that!*) Solver is an algorithm and works by incrementing its parameter values up to a maximum number of iterations. If we start it off too far away from the optimum solution, or point it in the wrong direction, then it may fail to converge, as it has here!

However, as the familiar adage advises, let's try again with different starter values. A simple look at the data would suggest that the project has started around 3 months late, so we can try using that as an alternative starting parameter. We can also try moving the Mode by the same amount from month 6 to month 9. Let's just increase the outturn randomly to € 6,000k, before trying again…

A word (or two) from the wise?

'If at first you don't succeed, Try, try, try again.'

William E Hickson
British Educational Writer
1803–1870

... and as if by magic, it now converges to the result in Table 2.19 and Figure 2.38 giving us a start month of 2.84, a Mode midway through month 10 (i.e. 9.581) and an outturn EAC of € 6,986 k.

If the false start has shaken your confidence in Solver, then try running it again several times using different input parameter values for start, mode and outturn. Hopefully this will return more consistent results, unless the actuals are not fitting a Norden-Rayleigh Curve pattern, in which case the transformation and regression are not valid.

The next question an enquiring estimator is bound to ask is 'How stable is this EAC?' Well, that's a good question! Let's add a couple of months' worth of data and re-run the model. We get the results in Table 2.20 and Figure 2.39, which give us a similar start position, but a slightly later mode and an increased EAC outturn.

In fact, if we were to track the EAC and projected completion date, we would find a pattern emerging, as shown in Table 2.21 and Figure 2.40. In this particular case we can see that as the schedule slips each month the cost outturn also increases at a rate greater than the Pro Rata Product Rule (Section 2.2.3) but less than the Square Rule (Section 2.2.4) and that the average of the two is a very close approximation to what has happened here. We can imply therefore that the cost outturn increases in line with a general quadratic relationship in relation to schedule slippage in this instance; we should not infer that this will always be the case and should not use this as justification for fitting a quadratic curve of best fit to the Norden-Rayleigh Cost Creep Trend Analysis in Figure 2.41; if we did it would turn back on itself!

Table 2.19 Solver Model Results for Linear Transformation of a Norden-Rayleigh Curve (1)

Parameter	Rayleigh Distribution
NRC Factor	3.5
Start	0
Mode	6
End	21
√2	1.414
alpha	2
beta	8.485
Total Budget	€ 5,000 k

Rayleigh Distribution	Time from Start
2.840	0
9.631	6.691
26.258	23.418
1.414	
2	
9.462	

Outturn > € 6,986 k

	Time from Start
Mode	6.691

Regression	
No of Obs	8
Slope	-0.011
Intercept	8.8538
RD-EAI	€ 7,001 k

Nominal End Month	Rayleigh Distribution CDF	Norden-Rayleigh Uplift
21.00	99.781%	100.219%

Outturn x NRC Uplift Factor: € 7,001 k (RD-EAI >)

Month	Rayleigh Distribution	NRC Budget	Actual Spend	Time from Start (Month - Start)	Best Fit Offset NRC	Error (Actual - Best Fit)	Time from Start Squared (s)	RD-ETI	Natural Log Ln(RD-ETI)	Best Fit Ln(RD-ETI)	Error (Best Fit - Ln(RD-ETI))
0	0.000%	€0 k	€0 k	-2.840	#N/A	#N/A	#N/A	7,001	8.8538	#N/A	#N/A
1	1.379%	€69 k	€0 k	-1.840	#N/A	#N/A	#N/A	7,001	8.8538	#N/A	#N/A
2	5.404%	€271 k	€0 k	-0.840	#N/A	#N/A	#N/A	7,001	8.8538	#N/A	#N/A
3	11.750%	€589 k	€12 k	0.160	€2 k	€10 k	0.03	6,989	8.8521	8.8535	-0.0014
4	19.926%	€998 k	€91 k	1.160	€104 k	-€13 k	1.35	6,910	8.8407	8.8388	0.0019
5	29.335%	€1,470 k	€346 k	2.160	€355 k	-€9 k	4.67	6,655	8.8031	8.8017	0.0014
6	39.347%	€1,972 k	€752 k	3.160	€739 k	€13 k	9.99	6,249	8.7402	8.7423	-0.0021
7	49.366%	€2,474 k	€1,234 k	4.160	€1,230 k	€4 k	17.30	5,767	8.6599	8.6605	-0.0006
8	58.889%	€2,951 k	€1,799 k	5.160	€1,801 k	-€2 k	26.62	5,202	8.5568	8.5564	0.0004
9	67.535%	€3,384 k	€2,415 k	6.160	€2,418 k	-€3 k	37.94	4,586	8.4307	8.4300	0.0007
10	75.065%	€3,761 k	€3,053 k	7.160	€3,052 k	€1 k	51.26	3,948	8.2809	8.2812	-0.0003
11	81.373%	€4,078 k		8.160	€3,673 k						
12	86.466%	€4,333 k		9.160	€4,258 k						Total Error: 0.0000
13	90.437%	€4,532 k		10.160	€4,791 k						SSE: 0.000013
14	93.427%	€4,682 k		11.160	€5,259 k						SSE x 10⁶:

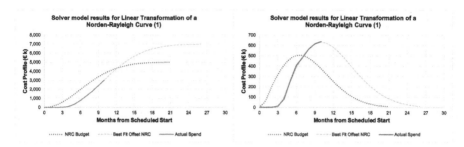

Figure 2.38 Solver Model Results for Linear Transformation of a Norden-Rayleigh Curve (1)

Table 2.20 Solver Model Results for Linear Transformation of a Norden-Rayleigh Curve (2)

Parameter	Rayleigh Distribution
NRC Factor	3.5
Start	0
Mode	6
End	21
√2	1.414
alpha	2
beta	8.485
Total Budget	€ 5,000 k

	Rayleigh Distribution	Time from Start
	-2.810	0
	9.712	6.902
	26.967	24.157
	1.414	
	2	
	9.761	

Outturn > € 7,273 k

Time from Start	
Mode	6.902

Nominal End Month	Rayleigh Distribution CDF	Norden-Rayleigh Uplift		Outturn x NRC Uplift Factor
21.00	99.781%	100.219%	RD-EAI >	€ 7,294 k

Regression	
No of Obs	10
Slope	-0.010
Intercept	8.8948
RD-EAI	€ 7,294 k

Month	Rayleigh Distribution	NRC Budget	Actual Spend	Time from Start (Month - Start)	Best Fit Offset NRC	Error (Actual - Best Fit)	Time from Start Squared (s)	RD-ETI	Natural Log Ln(RD-ETI)	Best Fit Ln(RD-ETI)	Error (Best Fit - Ln(RD-ETI))
0	0.000%	€ 0 k	€ 0 k	-2.810	#N/A	#N/A	#N/A	7,294	8.8948	#N/A	#N/A
1	1.379%	€ 69 k	€ 0 k	-1.810	#N/A	#N/A	#N/A	7,294	8.8948	#N/A	#N/A
2	5.404%	€ 271 k	€ 0 k	-0.810	#N/A	#N/A	#N/A	7,294	8.8948	#N/A	#N/A
3	11.750%	€ 589 k	€ 12 k	0.190	€ 3 k	€ 9 k	0.04	7,282	8.8931	8.8944	-0.0013
4	19.926%	€ 998 k	€ 91 k	1.190	€ 108 k	-€ 17 k	1.42	7,203	8.8822	8.8799	0.0023
5	29.335%	€ 1,470 k	€ 346 k	2.190	€ 358 k	-€ 12 k	4.80	6,948	8.8462	8.8444	0.0017
6	39.347%	€ 1,972 k	€ 752 k	3.190	€ 739 k	€ 13 k	10.18	6,542	8.7859	8.7879	-0.0020
7	49.366%	€ 2,474 k	€ 1,234 k	4.190	€ 1,227 k	€ 7 k	17.56	6,060	8.7094	8.7105	-0.0011
8	58.889%	€ 2,951 k	€ 1,799 k	5.190	€ 1,796 k	€ 3 k	26.94	5,495	8.6115	8.6120	-0.0005
9	67.535%	€ 3,384 k	€ 2,415 k	6.190	€ 2,415 k	€ 0 k	38.32	4,879	8.4926	8.4926	0.0000
10	75.065%	€ 3,761 k	€ 3,053 k	7.190	€ 3,054 k	-€ 1 k	51.70	4,241	8.3525	8.3522	0.0003
11	81.373%	€ 4,078 k	€ 3,683 k	8.190	€ 3,686 k	-€ 3 k	67.08	3,611	8.1916	8.1907	0.0009
12	86.466%	€ 4,333 k	€ 4,289 k	9.190	€ 4,288 k	€ 1 k	84.46	3,005	8.0079	8.0083	-0.0004
13	90.437%	€ 4,532 k		10.190	€ 4,841 k						
14	93.427%	€ 4,682 k		11.190	€ 5,334 k					Total Error	#N/A
15	95.606%	€ 4,791 k		12.190	€ 5,760 k					SSE	0.000016
16	97.143%	€ 4,868 k		13.190	€ 6,119 k					SSE x 10⁶	16

Now, as promised at the start of Section 2.6, we will return to the suggestion of Christensen and Rees (2002) that EVM systems can be used to predict EACs after they have reached the 20% completion level. This example shows that this may not be the case. The outturn prediction only starts to stabilise until we're safely past the Mode, which is at around 40% completion, and in this particular case we are in excess of 50% of the predicted trend! Let's just agree that it may be possible to get a stable view of a NRC EAC at a relatively early stage, but let's not bank on it; we should track it regularly.

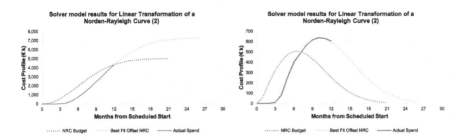

Figure 2.39 Solver Model Results for Linear Transformation of a Norden-Rayleigh Curve (2)

Table 2.21 Solver Outturn Creep Follows the Average of the Two Schedule Slippage Cost Rules

	Month	8	9	10	11	12	13
Solver Results	Count of Regression Points	5	7	8	9	10	11
	Effective Start	2.889	2.863	2.840	2.827	2.810	2.809
	Effective End	23.958	25.376	26.258	26.620	26.967	26.979
	Effective Mode	8.909	9.295	9.531	9.625	9.712	9.714
	Projected Outturn	€ 5,932 k	€ 6,588 k	€ 6,986 k	€ 7,140 k	€ 7,278 k	€ 7,281 k
	Projected Duration (End - Start)	21.069	22.513	23.418	23.793	24.158	24.170
	Duration Change as % of Month 8	100.0%	106.9%	111.1%	112.9%	114.7%	114.7%
	Schedule Slippage Pro Rata Cost Rule	€ 5,932	€ 6,339	€ 6,594	€ 6,699	€ 6,802	€ 6,805
	Schedule Slippage Square Cost Rule	€ 5,932	€ 6,773	€ 7,329	€ 7,566	€ 7,799	€ 7,807
	Average of Schedule Slippage Cost Rules	€ 5,932	€ 6,556	€ 6,961	€ 7,132	€ 7,300	€ 7,306

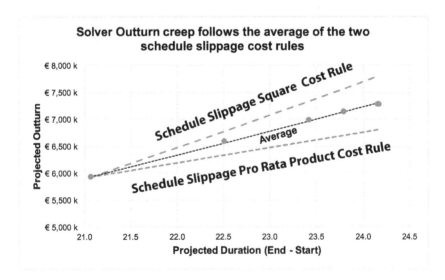

Figure 2.40 Solver Outturn Creep Follows the Average of the Two Schedule Slippage Cost Rules

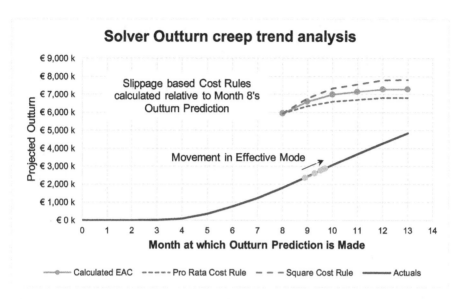

Figure 2.41 Solver Outturn Creep Trend Analysis

2.6.4 Exploiting Weibull Distribution's double log linearisation constraint

This technique is essentially a variation on that discussed in the last section, exploiting the same basic property.

If we look back at Volume II Chapter 4 on the properties of Weibull Distributions we will find that we can transform any Weibull Distribution into a Linear Function using a double Natural Log. As a Norden-Rayleigh Curve is based on the Rayleigh Distribution which is just a special case of a Weibull Distribution, then the same must be true of a NRC.

For the Formula-philes: Alternative linear representation of a Norden-Rayleigh Curve

Consider a Norden-Rayleigh Curve generated from a Rayleigh Distribution with a Start point at time 0, a Mode at time λ and a nominal completion time at time τ, giving an uplift factor of k relative to the corresponding Rayleigh Distribution, and an Outturn Cost of C_T

(Continued)

Cumulative Norden-Rayleigh Curve $N_\lambda(t)$ at time t:

$$N_\lambda(t) = kC_T\left(1 - e^{-\frac{1}{2}\left(\frac{t}{\lambda}\right)^2}\right)$$ (1)

Simplifying (1) to give the percentage of work outstanding:

$$1 - \frac{N_\lambda(t)}{kC_T} = e^{-\frac{1}{2}\left(\frac{t}{\lambda}\right)^2}$$ (2)

Taking the Natural Log of (2), substituting Ln(e) = 1, and then re-arranging:

$$\ln\left(1 - \frac{N_\lambda(t)}{kC_T}\right) = -\frac{1}{2}\left(\frac{t}{\lambda}\right)^2$$ (3)

As the percentage to completion is less than 1, the log is less than 0, we need to reverse the signs:

$$-\ln\left(1 - \frac{N_\lambda(t)}{kC_T}\right) = \frac{1}{2}\left(\frac{t}{\lambda}\right)^2$$ (4)

Taking the Natural Log of (4):

$$\ln\left(-\ln\left(1 - \frac{N_\lambda(t)}{kC_T}\right)\right) = \ln\left(\frac{t}{\sqrt{2\lambda}}\right)^2$$ (5)

Simplifying (5):

$$\ln\left(-\ln\left(1 - \frac{N_\lambda(t)}{kC_T}\right)\right) = 2\ln t - 2\ln\left(\sqrt{2\lambda}\right)$$ (6)

... which is the equation of a straight line with a slope of 2 and an intercept of $-2\ln\left(\sqrt{2\lambda}\right)$

The relationship can be expressed as the:

Double Natural Log of the Underpinning Rayleigh Distribution's Cumulative Percentage to Completion is a linear function of the **Natural Log of the Elapsed Time**

... *which admittedly doesn't quite flow off the tongue too readily.*

In other words, this is another of those '*we need to know the answer, to get the answer*', but as we have already seen in the previous section, Microsoft Excel's Solver can help us to get around that one.

It has the property that the slope of this transformation is always 2, and that the Intercept is always -2 LN(Mode $\sqrt{2}$).

Table 2.22 and Figure 2.42 illustrate the transformation with a 'perfect' Norden-Rayleigh Curve:

- The outstanding Rayleigh Distribution % is 100% minus Column (A)
- Column (F) is the reciprocal of Column (E)
- Column (G) is the Natural Log of the Natural Log of Column (F), i.e. LN(LN(Column F))

Table 2.22 Transformation of a 'Perfect' Norden–Rayleigh Curve to a Linear Form

Rayleigh Distribution	
NRC Factor	3.5
Start	0
Mode	6
End	21.00
√2	1.414
alpha	2
beta	8.485
Total Budget	€ 5,000 k

	Linear Transformation Intercept

(a)	(b)	(c)	(d)
Ln(Mode)	Ln(√2)	(a) + (b)	-2 (c)
1.792	0.3466	2.1383	-4.2767

Time	Rayleigh Distribution CDF	Norden-Rayleigh Uplift
21.00	99.781%	100.219%

(A)	(B)	(C)	(D)	(E)	(F)	(G)
Month	Rayleigh Distribution	NRC Budget (Ref Only)	Natural Log of Month	Outstanding Rayleigh %	Reciprocal of Outstanding Rayleigh %	Double Natural Log of Reciprocal of Outstanding Rayleigh %
1	1.38%	€ 69 k	0.000	98.62%	1.0140	-4.2767
2	5.40%	€ 271 k	0.693	94.60%	1.0571	-2.8904
3	11.75%	€ 589 k	1.099	88.25%	1.1331	-2.0794
4	19.93%	€ 998 k	1.386	80.07%	1.2488	-1.5041
5	29.34%	€ 1,470 k	1.609	70.66%	1.4151	-1.0578
6	39.35%	€ 1,972 k	1.792	60.65%	1.6487	-0.6931
7	49.37%	€ 2,474 k	1.946	50.63%	1.9750	-0.3848
8	58.89%	€ 2,951 k	2.079	41.11%	2.4324	-0.1178
9	67.53%	€ 3,384 k	2.197	32.47%	3.0802	0.1178
10	75.06%	€ 3,761 k	2.303	24.94%	4.0104	0.3285
11	81.37%	€ 4,078 k	2.398	18.63%	5.3685	0.5191
12	86.47%	€ 4,333 k	2.485	13.53%	7.3891	0.6931
13	90.44%	€ 4,532 k	2.565	9.56%	10.4565	0.8532
14	93.43%	€ 4,682 k	2.639	6.57%	15.2141	1.0014
15	95.61%	€ 4,791 k	2.708	4.39%	22.7599	1.1394
16	97.14%	€ 4,868 k	2.773	2.86%	35.0073	1.2685
17	98.19%	€ 4,920 k	2.833	1.81%	55.3617	1.3898
18	98.89%	€ 4,955 k	2.890	1.11%	90.0171	1.5041
19	99.34%	€ 4,978 k	2.944	0.66%	150.4888	1.6122
20	99.61%	€ 4,992 k	2.996	0.39%	258.6706	1.7148
21	99.78%	€ 5,000 k	3.045	0.22%	457.1447	1.8124

Slope (G-range,D-range)	2.0000
Intercept (G-range,D-range)	-4.2767

The slope of the transformation is always 2, and the intercept is always $-2\,\mathrm{LN}(\mathrm{Mode}\,\sqrt{2})$.

However, by way of verification that the double Log Transformation is working, we have also shown these values calculated using the Excel functions **SLOPE** and **INTERCEPT** using range (G) as the *known-y's* and range (D) as the *known-x's*.

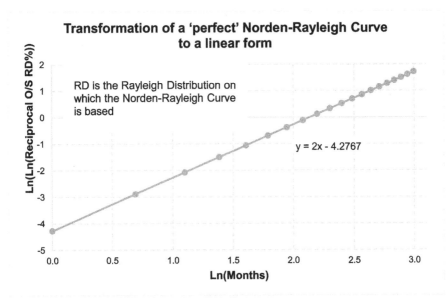

Figure 2.42 Transformation of a 'Perfect' Norden-Rayleigh Curve to a Linear Form

Now let's try exploiting this property using our example actual data we considered in the previous three techniques. We are going to use Solver to find the Best Fit Norden-Rayleigh Curve to the actual data, but with the added constraint that exploits this Double Log linear transformation property. Table 2.23 illustrates an appropriate set-up procedure.

1. The left-hand three columns set-up the Budget Norden-Rayleigh Curve as a reference
2. The Actual costs to date are in the fourth column
3. The fifth column header holds the parameter values for our intended 'Best Fit' NRC model. The Start point, Mode and Outturn Cost (EAC) will be used as the Solver variables but initially can be set at random. Here we have assumed that they are the Budget values
4. The lower section of the fifth column offsets the Time (Months) by the Start point value (initially zero offset)
5. The sixth column header summarises the revised Start and Mode parameters that will be used in the revised NRC calculation (i.e. they are normalised to reflect a start time of zero). It also multiplies the Outturn parameter by the Uplift Factor we have chosen to apply. Here we are assuming the 3.5 Truncation Ratio discussed previously. This gives us the theoretical cost at infinity required for the underlying Rayleigh Distribution which we will use in Steps 6 and 9

Table 2.23 Solver Setup Exploiting the Linear Transformation Property of a Norden-Rayleigh Curve

Budget Rayleigh Distribution				Best Fit Rayleigh Distribution	Time from Start					
NRC Factor	3.5									
Start	0			0.000	0				Linear Transformation Intercept	
Mode	6	0.00		6.000	3.049					
End	21.00			21.000	10.670		(a)	(b)	(c)	(d)
√2	1.414			1.414	1.414		Ln(Mode)	Ln(√2)	(a)+(b)	-2 (c)
alpha	2			2	2		1.115	0.3466	1.4613	-2.9225
beta	8.485			8.485	4.311					
Total Budget	€ 5,000 k		Outturn >	€ 5,000 k						

Time	Rayleigh Distribution CDF	Norden-Rayleigh Uplift
21.00	99.781%	100.219%

(A) Month	(B) Rayleigh Distribution	(C) Norden-Rayleigh Budget	Actual Spend	Time from Start: Month - Start	Current NRC Model	Error	(D) Natural Log of Time from Start	(E) Outstanding Rayleigh %	(F) Reciprocal of Outstanding Rayleigh %	(G) Double Natural Log of Reciprocal of Outstanding Rayleigh %
0	0.00%	€ 0 k	€ 0 k	0.000	€ 0 k	€ 0 k	#NUM!	100.00%	1.0000	#N/A
1	1.38%	€ 69 k	€ 0 k	1.000	€ 262 k	-€ 262 k	0.000	100.00%	1.0000	#N/A
2	5.40%	€ 271 k	€ 0 k	2.000	€ 970 k	-€ 970 k	0.693	100.00%	1.0000	#N/A
3	11.75%	€ 589 k	€ 12 k	3.000	€ 1,923 k	-€ 1,911 k	1.099	99.81%	1.0019	-6.2473
4	19.93%	€ 998 k	€ 91 k	4.000	€ 2,892 k	-€ 2,801 k	1.386	98.53%	1.0149	-4.2149
5	29.34%	€ 1,470 k	€ 346 k	5.000	€ 3,705 k	-€ 3,359 k	1.609	94.42%	1.0590	-2.8582
6	39.35%	€ 1,972 k	€ 752 k	6.000	€ 4,288 k	-€ 3,536 k	1.792	87.88%	1.1379	-2.0466
7	49.37%	€ 2,474 k	€ 1,234 k	7.000	€ 4,652 k	-€ 3,418 k	1.946	80.11%	1.2482	-1.5064
8	58.89%	€ 2,951 k	€ 1,799 k	8.000	€ 4,851 k	-€ 3,052 k	2.079	71.01%	1.4083	-1.0719
9	67.53%	€ 3,384 k	€ 2,415 k	9.000	€ 4,947 k	-€ 2,532 k	2.197	61.08%	1.6371	-0.7074
10	75.06%	€ 3,761 k	€ 3,053 k	10.000	€ 4,988 k	-€ 1,935 k	2.303	50.80%	1.9685	-0.3897

				Sum of Errors (Ref only)	-€ 23,777 k	Regression Points		Slope	4.728
				Sum of Squares Error	87449499	8		Intercept	-10.8981

6. The lower section of the sixth column depicts the revised NRC as defined above. Initially, as we have assumed the budget parameters, this column is identical to the third column (although it won't be when Solver starts to run)

7. The seventh column calculates the error between the Actual Cumulative Cost and the revised NRC. It summarises the Sum of Squares Error (or SSE); this is the value which we will get Solver to minimise, i.e. our usual Least Squares technique

8. The last four columns are a repeat of the logic used in Table 2.22. In the header section we have calculated the Linear Transformation Intercept based on the revised Mode as -2 multiplied by the natural log of the revised mode and the square root of 2.

9. In the lower section of these four columns we perform the linear transformation of the revised NRC using the double natural log of the reciprocal of the outstanding Rayleigh Distribution percentage. The important thing to note here is that the outstanding Rayleigh Distribution percentage in the column marked (E) is calculated by taking the Outturn (€ 5,000 k as the starting value in this case) minus the Actuals divided by the Outturn multiplied by the uplift factor. (*Although in practice it makes little difference to the result*)

10. We can then determine the number of data points that we should use in our regression calculation of the slope and intercept of this linear transformation. If the revised start time is less than 1, we have chosen to exclude this from the regression. Values less than

1 generate negative natural log values, usually associated with a comparatively small cumulative cost. This in turn can create a distorted view of the transformation resulting in wildly inflated or deflated outturn values, as shown in Figure 2.43; this particular fit gives an EAC in excess of € 13,000 k! (*Now you might think that this 'random' exclusion sounds very naughty but it is done from a pragmatic perspective, but I note your concern*)

- Depending on the value of the Start point parameter the number of Regression Points to include will change dynamically within Solver. To avoid a failure to converge error, the number of Regression Points to include can be determined by taking the minimum of **COUNTIF**(*range(D)*,'>0') and **COUNTIF**(*range(F)*,'>1'). The latter avoids an error value in column denoted as (G) by taking the Double Natural Logarithm of 1 as **LN(1)=0**, so **LN(LN(1))** gives an **#NUM** error

11. Next we calculate the slope and intercept of the linear transformation using the selected data points and the Microsoft Excel functions **SLOPE**(*y-range, x-range*) and **INTERCEPT**(*y-range, x-range*) where the *y-range* is in the column denoted as (G) and the *x-range* is in that denoted as (D)

- As the number of Regression Points to include will vary according to Step 10, we can use the **OFFSET** function in conjunction with **SLOPE** and **INTERCEPT** for the *y-range* and *x-range*

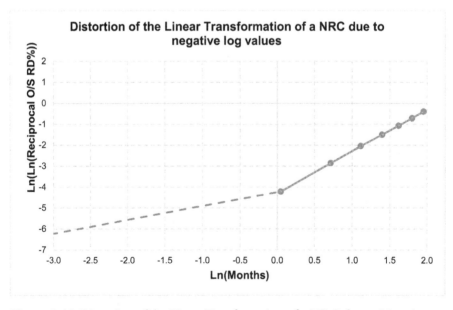

Figure 2.43 Distortion of the Linear Transformation of a NRC due to Negative Log Values

- **y-range = OFFSET(G-Blank, -Regression Points, 0, Regression Points, 1)**
  where G-Blank is the blank cell below the last data point in column denoted (G)
- **x-range = OFFSET(D-Blank, -Regression Points, 0, Regression Points, 1)** where
  D-Blank is the blank cell below the last data point in column denoted (D)
- The **OFFSET** function then counts back the number of non-negative data
  points from Step 10 and creates an array or range one column wide and includes the number of rows defined by the number of Regression Points

12. We can now set up the Excel Solver to minimise the Sum of Squares Error value
    subject to the constraints that the Slope must be 2 (initially 4.8120), and the two
    intercept calculations must be equal also (initially they are 4.814 and -10.7986.) It
    is also a good idea to tell Solver that the Endpoint must be greater than 'time now'
    and that the outturn must be greater than the Actual cost to date, just to avoid it
    scooting off in the wrong direction and failing to converge
13. Finally, let's not forget to uncheck the box that forces all unconstrained parameters
    to be non-negative!

Clicking 'Solve' will yield us results similar to those in Table 2.24. This gives
us an Outturn or EAC of some € 6,205 k and a completion date just after the
end of Month 24, having started just before the end of month 3, and reaching a
Mode at month 9 (and a bit). Figure 2.44 illustrates the result. It appears to be a

Table 2.24 Solver Results Exploiting the Linear Transformation Property of a Norden-Rayleigh Curve (1)

Budget Rayleigh Distribution		Best Fit Rayleigh Distribution	Time from Start
NRC Factor	3.5		
Start	0	2.951	0
Mode	6	0.00	9.019	8.068
End	21.00	24.189	21.238
√2	1.414	1.414	1.414
alpha	2	2	2
beta	8.485	8.581	8.581
Total Budget	€ 5,000 k	Outturn > € 6,205 k	

Linear Transformation Intercept

(a)	(b)	(c)	(d)
Ln(Mode)	Ln(√2)	(a) + (b)	-2 (c)
1.803	0.3466	2.1496	-4.2992

Time	Rayleigh Distribution CDF	Norden-Rayleigh Uplift
21.00	99.781%	100.219%

(A)	(B)	(C)					(D)	(E)	(F)	(G)
Month	Rayleigh Distribution	Norden-Rayleigh Budget	Actual Spend	Time from Start: Month - Start	Current NRC Model	Error	Natural Log of Time from Start	Outstanding Rayleigh %	Reciprocal of Outstanding Rayleigh %	Double Natural Log of Reciprocal of Outstanding Rayleigh %
0	0.00%	€ 0 k	€ 0 k	-2.951	#N/A	€ 0 k	#NUM!	100.00%	1.0000	#N/A
1	1.38%	€ 69 k	€ 0 k	-1.951	#N/A	€ 0 k	#NUM!	100.00%	1.0000	#N/A
2	5.40%	€ 271 k	€ 0 k	-0.951	#N/A	€ 0 k	#NUM!	100.00%	1.0000	#N/A
3	11.75%	€ 589 k	€ 12 k	0.049	€ 0 k	€ 12 k	-3.023	99.81%	1.0019	-6.2473
4	19.93%	€ 998 k	€ 91 k	1.049	€ 92 k	-€ 1 k	0.048	98.53%	1.0149	-4.2149
5	29.34%	€ 1,470 k	€ 346 k	2.049	€ 345 k	€ 1 k	0.717	94.42%	1.0590	-2.8582
6	39.35%	€ 1,972 k	€ 752 k	3.049	€ 737 k	€ 15 k	1.115	87.88%	1.1379	-2.0466
7	49.37%	€ 2,474 k	€ 1,234 k	4.049	€ 1,241 k	-€ 7 k	1.398	80.11%	1.2482	-1.5064
8	58.89%	€ 2,951 k	€ 1,799 k	5.049	€ 1,820 k	-€ 21 k	1.619	71.01%	1.4083	-1.0719
9	67.53%	€ 3,384 k	€ 2,415 k	6.049	€ 2,435 k	-€ 20 k	1.800	61.08%	1.6371	-0.7074
10	75.06%	€ 3,761 k	€ 3,053 k	7.049	€ 3,052 k	€ 1 k	1.953	50.80%	1.9685	-0.3897

Sum of Errors (Ref only)	-€ 20 k	Regression Points		Slope	2.000
Sum of Squares Error		7		Intercept	-4.2992

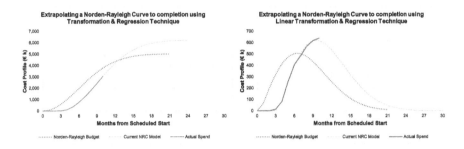

Figure 2.44 Extrapolating a NRC to Completion Using a Transformation and Regression Technique (1)

Table 2.25 Solver Results Exploiting the Linear Transformation Property of a Norden-Rayleigh Curve (2)

Budget Rayleigh Distribution				Best Fit Rayleigh Distribution	Time from Start					
NRC Factor	3.5								Linear Transformation Intercept	
Start	0			2.930	0					
Mode	6	0.00		9.390	6.461					
End	21.00			25.542	22.613					
√2	1.414			1.414	1.414		(a)	(b)	(c)	(d)
alpha	2			2	2		Ln(Mode)	Ln(√2)	(a) + (b)	-2 (c)
beta	8.485			9.137	9.137		1.866	0.3466	2.2123	-4.4246
Total Budget	€ 5,000 k		Outturn >	€ 6,826 k						

	Rayleigh Distribution CDF	Norden-Rayleigh Uplift								
Time										
21.00	99.781%	100.219%								

(A)	(B)	(C)					(D)	(E)	(F)	(G)
Month	Rayleigh Distribution	Norden-Rayleigh Budget	Actual Spend	Time from Start: Month - Start	Current NRC Model	Error	Natural Log of Time from Start	Outstanding Rayleigh %	Reciprocal of Outstanding Rayleigh %	Double Natural Log of Reciprocal of Outstanding Rayleigh %
0	0.00%	€ 0 k	€ 0 k	-2.930	#N/A	€ 0 k	#NUM!	100.00%	1.0000	#N/A
1	1.38%	€ 69 k	€ 0 k	-1.930	#N/A	€ 0 k	#NUM!	100.00%	1.0000	#N/A
2	5.40%	€ 271 k	€ 0 k	-0.930	#N/A	€ 0 k	#NUM!	100.00%	1.0000	#N/A
3	11.75%	€ 589 k	€ 12 k	0.070	€ 0 k	€ 12 k	-2.653	99.82%	1.0018	-6.3426
4	19.93%	€ 998 k	€ 91 k	1.070	€ 93 k	-€ 2 k	0.068	98.67%	1.0135	-4.3109
5	29.34%	€ 1,470 k	€ 346 k	2.070	€ 342 k	€ 4 k	0.728	94.93%	1.0534	-2.9561
6	39.35%	€ 1,972 k	€ 752 k	3.070	€ 730 k	€ 22 k	1.122	88.98%	1.1238	-2.1479
7	49.37%	€ 2,474 k	€ 1,234 k	4.070	€ 1,231 k	€ 3 k	1.404	81.92%	1.2207	-1.8124
8	58.89%	€ 2,951 k	€ 1,799 k	5.070	€ 1,813 k	-€ 14 k	1.623	73.64%	1.3579	-1.1844
9	67.53%	€ 3,384 k	€ 2,415 k	6.070	€ 2,441 k	-€ 26 k	1.803	64.62%	1.5476	-0.8286
10	75.06%	€ 3,761 k	€ 3,053 k	7.070	€ 3,082 k	-€ 29 k	1.956	55.27%	1.8093	-0.5227
11	81.37%	€ 4,078 k	€ 3,683 k	8.070	€ 3,705 k	-€ 22 k	2.088	46.04%	2.1720	-0.2541
12	86.47%	€ 4,333 k	€ 4,289 k	9.070	€ 4,287 k	€ 2 k	2.205	37.16%	2.6909	-0.0102

				Sum of Errors (Ref only)	-€ 53 k		Regression Points		Slope	2.000
				Sum of Squares Error	2344		9		Intercept	-4.4246

89506

reasonably good cumulative fit (left-hand graph) but the right-hand graph depicting the monthly spend suggests that the Modal value has been understated, relatively speaking.

If we re-run the model two months later as we did with the basic Least Squares Solver technique in the last section, we again get a different result, as shown in Table 2.25 and

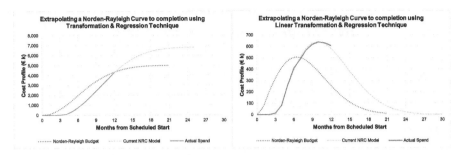

Figure 2.45 Extrapolating a NRC to Completion Using a Transformation and Regression Technique (2)

Table 2.26 Solver Outturn Creep Follows a Weighted Average of the Two Schedule Slippage Cost Rules

	Month	8	9	10	11	12	13
Solver Results	Count of Regression Points	5	6	7	8	9	10
	Effective Start	2.984	2.966	2.951	2.940	2.930	2.922
	Effective End	21.351	23.013	24.189	24.959	25.542	25.900
	Effective Mode	8.232	8.694	9.019	9.231	9.390	9.487
	Projected Outturn	€ 4,895 k	€ 5,657 k	€ 6,205 k	€ 6,562 k	€ 6,826 k	€ 6,981 k
	Projected Duration (End - Start)	18.367	20.047	21.238	22.019	22.613	22.978
Weighting	Duration Change as % of Month 8	100.0%	109.1%	115.6%	119.9%	123.1%	125.1%
45%	Schedule Slippage Pro Rata Cost Rule	€ 4,895	€ 5,343	€ 5,660	€ 5,869	€ 6,027	€ 6,124
55%	Schedule Slippage Square Cost Rule	€ 4,895	€ 5,832	€ 6,545	€ 7,036	€ 7,420	€ 7,662
	Weighted Average of Schedule Slippage Cost Rules	€ 4,895	€ 5,612	€ 6,147	€ 6,510	€ 6,793	€ 6,970

Figure 2.45. With an Outturn of € 6,826 k with a completion date midway through Month 26, having commenced just before the end of month 3. Worryingly, however, these values are significantly lower than those generated by the previous technique; also, we will note that the right-hand graph of Figure 2.45 suggests that the programme is likely to slip even more with a knock-on to costs. Whilst this technique exploits a property of the Rayleigh Distribution, it does pre-suppose that the model is true in an absolute sense, which may be a somewhat misplaced reliance. (*After all, since when did anything to do with estimating become an exact science?*)

In Table 2.26 and Figure 2.46 we can track the performance of this technique over time and there appears to be little convergence to a steady state condition. This technique may suffer from an over reliance on a model which is really only an empirical observation that employs a truncation rule in order to work. The mere fact that we had to interject a fairly random step of ignoring negative Log Values should have warned us that pragmatic decisions are not always practical decisions.

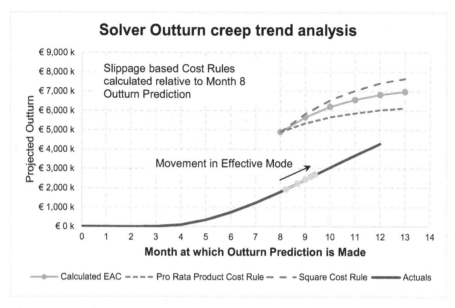

Figure 2.46 Solver Outturn Creep Trend Analysis

Our conclusion here must surely be that this may not be the most robust technique (or application of it). It seems to be a question of a good theory looking for a relevant problem to solve. *Perhaps this is one to put in the 'Don't try this at home, or at work' pile* or may be one just to use as a sensitivity check value. In other situations, it may work a treat.

2.6.5 Estimates to Completion – Review and conclusion

As a refresher let's compare the results of Period 10 and Period 12 across the four techniques and six answers. (*We have two techniques that can be used with either the Norden-Rayleigh Curve or the PERT-Beta Lookalike.*)

Table 2.27 summarises the results, and as we can see, there is a degree of variation in the values generated. Before we throw it all in the bin, let's look on the bright side; we can use the various techniques to develop a range estimate as shown in Figure 2.47. If we were to continue the process with additional monthly updates, we would find (*unsurprisingly*) that the set of technique results are likely to continue to converge; this is a typical example of the Cone of Uncertainty (Volume I Chapter 4) articulated by Bauman (1958) and Boehm (1981) applied to the life of an Estimate to

Table 2.27 Comparison of Outturn Predictions Using Four Techniques

Month 10 Results			Start	Mode	End	Duration	Outturn	Comment
Guess & Iterate	NRC	Range for	3	10.36	30	27.00	€ 8,250 k	Hit & Miss Technique with
	PERT-Beta	either	3	9	24	21.00	€ 6,124 k	either model
Basic Least Squares	NRC		2.86	9.40	25.74	22.87	€ 6,777 k	Optimistic
Excel Solver	PERT-Beta		2.88	9.92	27.51	24.63	€ 7,697 k	Pessimistic
Linear Transformation & Regression			2.84	9.53	26.26	23.42	€ 6,986 k	Realistic?
Weibull Double Natural Log Linearisation			2.95	9.02	24.19	21.24	€ 6,205 k	Overly precise constraint
			Fairly consistent	Degree of variation in the Mode with consequential impact on Endpoint (3.5 ratio rule)			> 99% Correlation with both the Mode and the Duration	

Month 12 Results		Start	Mode	End	Duration	Outturn	Comment
Basic Least Squares	NRC	2.83	9.66	26.73	23.90	€ 7,192 k	Optimistic
Excel Solver	PERT-Beta	2.89	9.86	27.28	24.39	€ 7,602 k	Pessimistic
Linear Transformation & Regression		2.81	9.71	26.97	24.16	€ 7,278 k	Realistic?
Weibull Double Natural Log Linearisation		2.93	9.39	25.54	22.61	€ 6,826 k	Overly precise constraint
		Fairly consistent	Reduced variation in the Mode compared with Month 10 with consequential impact on Endpoint			99% Correlation with the Mode and 94% Correlation with the Duration	

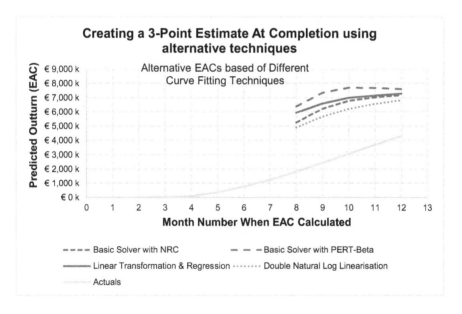

Figure 2.47 Creating a 3-Point Estimate at Completion Using Alternative Techniques

Completion. Here, we are progressively replacing and reducing uncertainty around future values with the certainty of actuals. The key thing for us here is to get over the hump, i.e. the mode (*rather than get the hump over the differences*). In this test, the PERT-Beta appears to be more likely to continue to give a slightly more pessimistic forecast, but generally appears to stabilise earlier, whereas the Double Natural Log

Linearisation still tends to be understated. A combination of these techniques could be used to generate a 3-Point Estimate.

We mustn't lose sight of the fact that the Norden-Rayleigh Curve is an empirical relationship, around which certain development behavioural characteristics have been postulated. If those characteristics do not describe our situation then we may find that none of these EAC/ETC techniques are appropriate. However, we may be able to use the general Least Squares Curve Fitting technique described in Volume III using either a general Weibull or, perhaps even better to avoid truncation issues, a general Beta Distribution. The Double Natural Log Linearisation Technique could be adapted for a general Weibull Distribution, but as we have seen with the Norden-Rayleigh Curve, it may be trying to enforce too exacting a constraint on what is after all only a representation of reality and not a physical law.

2.7 Chapter review

In this chapter we have looked at using Norden-Rayleigh Curves for Non-Recurring Development (or NRC_4NRD).

We started by briefly reviewing its history and how Norden observed that spend patterns on Non-Recurring Development projects followed a Rayleigh Distribution. He postulated reasons why this was the case and expressed a number of conditions in which the empirical relationship was valid; the main one was that the development should be a self-contained development for which the final objective remains unchanged.

The main problem with a Rayleigh Distribution is that it goes on forever (*sorry, did someone say, 'Just like this chapter'?*) Although some development projects may seem to be never-ending, we do need to have a practical limit on when the development is deemed to have been completed. There is no single view of when this should be; some choose the 97% Confidence Level as the truncation point giving a 2.65 ratio between the development Mode (peak effort) and the endpoint. Others prefer to use a 3.5:1 ratio between Mode and endpoint. The benefit of the latter is that it requires much less 'adjustment' to take account of premature truncation.

For those of us who are averse to random truncations, we can always consider using a PERT-Beta Distribution with parameters $\alpha = 2^1/_7$ and $\beta = 3^6/_7$. (*Note that 7 is twice the NRC Truncation Ratio of 3.5, which may make it easier to remember.*)

Whilst the NRC was originally conceived for projects that follow a natural development cycle unhindered, we can assess the impact when these 'rules are broken'. This gives us some rule of thumb for schedule slippage such as the Pro Rata Product Cost Rule and the Square Cost Rule, depending on the reason for the slippage. We can also model phased developments by layering multiple Norden-Rayleigh Curves. Depending on the level and timing of the additional development tasks, we can get a Dromedary or Bactrian Camel effect. This 'camelling' effect might be suitably modelled as general Weibull or Beta Distributions. We might find also that in certain circumstances a phased development project may easily be mistaken for a pure Norden-Rayleigh single contained development project. In the scheme of things, it probably does not matter.

We ended our exploration into Norden-Rayleigh Curves by looking at potential techniques we can try to generate Estimates to Completion for existing Development projects. We could use a 'Guess and Iterate' technique, but that is prone to subjective opinion and fails our TRACEability paradigm. We could use a Least Squares Curve Fitting with either the Norden-Rayleigh Curve or the PERT-Beta Lookalike, the latter potentially giving a more pessimistic perspective than the NRC option.

If we exploit the properties of a Rayleigh Distribution (as a special case of the Weibull Distribution) and transform the Estimate To Completion into a linear format, we can then perform a linear regression. However, if we go for the Double Natural Log transformation of a Weibull Distribution then it appears to place too much of a purist constraint on what is after all an empirical relationship. As a consequence, this last technique may give spurious results until later in the project life cycle, and so is probably not the best option to use.

All this makes you wonder if the effort required by an organisation to pull together a major complex bid for new work, also follows a Norden-Rayleigh Curve? Now, who's developing a headache? Time for a lie-down.

References

Amneus, D & Hawken, P (1999) *Natural Capitalism: Creating the Next Industrial Revolution*, Boston, Little Brown & Co, p.272.

Augustine, NR (1997) *Augustine's Laws (6th Edition)*, Reston, American Institute of Aeronautics and Astronautics, Inc.

Bauman, HC (1958) 'Accuracy considerations for capital cost estimation', *Industrial & Engineering Chemistry*, April.

Boehm, BW (1981) *Software Engineering Economics*, Upper Saddle River, Prentice-Hall.

Christensen, DS & Rees, DA (2002) 'Is the CPI-based EAC a lower bound to the final cost of post A-12 contracts?', *The Journal of Cost Analysis and Management*, Winter 2002.

Dukovich, J, Houser, S & Lee, DA (1999) *The Rayleigh Analyzer: Volume 1 – Theory and Applications, AT902C1*, McLean, Logistics Management Institute.

Gallagher, MA & Lee, DA (1996) 'Final-Cost Estimates for research & development programs conditioned on realized costs', *Military Operations Research*, Volume 2, Number 2: pp.51–65.

Fazar, W (1959) 'Program evaluation and review technique', *The American Statistician*, Volume 13, Number 2, April: p.10.

Lee, DA, Hogue, MR & Hoffman, DC (1993) *Time Histories of Expenditures for Defense Acquisition Programs in the Development Phase. – Norden-Rayleigh and Other Models*, ISPA Annual Meeting.

Lee, DA (2002) 'Norden-Raleigh Analysis: A useful tool for EVM in development projects', *The Measurable News*, Logistics Management Institute, March.

Norden, PV (1963) 'Useful tools for project management' in Dean, BV (Ed.) *Operations Research in Research and Development*, John Wiley and Sons.

Putnam, LH & Myers, W (1992) *Measures for Excellence – Reliable Software on Time, within Budget,* Upper Saddle River, Prentice-Hall.

Younossi, O, Stem, DE, Lorell, MA & Lussier, FM (2005) *Lessons Learned from the F/A-22 and F/A-18 E/F Development Programs*, Santa Monica, CA, RAND Corporation, [online] Available from: http://www.rand.org/pubs/monographs/MG276.html [Accessed 24–01–2017].

3

Monte Carlo Simulation and other random thoughts

Many seasoned estimating practitioners will tell you that a Range Estimate is always better than a single-point deterministic estimate. (*We have a better chance of being right, or less chance of being wrong, if we are one of those 'glass is half-empty' people!*)

If we create a Range Estimate (or 3-Point Estimate) using Monte Carlo Simulation we are in effect estimating with random numbers (*... and I've lost count of the number of Project Managers who have said, 'Isn't that how you estimators usually do it?'*).

There is, of course, a theory and structure to support the use of Monte Carlo Simulation; after all, as Robert Coveyou is reported to have commented (Peterson, 1997), we wouldn't want its output to be a completely chance encounter, would we?

> ### A word (or two) from the wise?
>
> *'The generation of random numbers is too important to be left to chance.'*
> **Robert R. Coveyou**
> American Research
> Mathematician
> Oak Ridge National Laboratory
> 1915–1996

3.1 Monte Carlo Simulation: Who, what, why, where, when and how

3.1.1 Origins of Monte Carlo Simulation: Myth and mirth

Based purely on its name, it is often assumed that the technique was invented by, or for use in, the gambling industry in Monte Carlo in order to minimise the odds of the 'house' losing, except on an occasional chance basis ... after all someone's good luck is their good public relations publicity. However, there is a link, albeit a somewhat more tenuous one, between Monte Carlo Simulation and gambling.

It was 'invented' as a viable numerical technique as part of the Manhattan Project which was an international research programme tasked with the development of nuclear weapons by the USA, UK and Canada (Metropolis & Ulam, 1949). (*So why, wasn't it called 'Manhattan Simulation' or the 'Big Bang Theory Simulation'? You may well ask … but it just wasn't.*) The phrase 'Monte Carlo Simulation' was coined as a codename by Nicholas Metropolis in recognition that the uncle of its inventor, colleague Stanislaw Ulam, used to frequent the casinos in Monaco hoping to chance his luck. (*The name 'Metropolis' always conjures up images of Superman for me, but maybe that's just the comic in me wanting to come out? In this context we could regard Stanislaw Ulam as a super hero.*)

The Manhattan Project needed a repeatable mathematical model to solve complex differential equations that could not be solved by conventional deterministic mathematical techniques. Monte Carlo Simulation gave them a viable probabilistic technique. The rest is probably history.

The fundamental principle behind Monte Carlo Simulation is that if we can describe each input variable to a system or scenario by a probability distribution (i.e. the probability that the variable takes a particular value in a specified range) then we can model the likely outcome of several independent or dependent variables acting together.

3.1.2 Relevance to estimators and planners

Let me count the ways 1, 2, 3, 4 …. Err, 12, 13, 14, 15… dah! I've lost count, but I think I got up to 'umpteen'.

It has been used extensively in the preparation of this book to demonstrate likely outcomes of certain worked examples and empirical results.

We can use it to model:

- The range of possible outcomes, and their probabilities, of particular events or combination of events where the mathematics could be computed by someone that way inclined, … and not all estimators have or indeed need that skill or knowledge
- The range of possible outcomes, and their probabilities, of particular events or combination of events, where the mathematics are either nigh on impossible to compute, or are just totally impractical from a time perspective, even for an estimator endowed with those advanced mathematical skills. (*Why swim the English Channel when you can catch a ferry or a train? It may not feel as rewarding but it's a lot quicker and probably less risky too.*) The original use in the Manhattan Project is the prime example of this
- We can use it to model assumptions to test out how likely or realistic certain technical or programmatic assumptions might be, or how sensitive their impact may be for cost or schedule, for example
- … and, of course, to model risk, opportunity and uncertainty in cost and/or schedule to generate 3-Point Estimates of Cost Outturn and Forecast Completion Dates, which definitely falls into the second group above

Over the next few sub-sections, we'll take a look at a few examples to demonstrate its potential use before we delve deeper into cost and schedule variability in the rest of the chapter.

Note: Whilst we can build and run a Monte Carlo Model in Microsoft Excel, it is often only suitable for relatively simple models with a limited number of independent input variables. Even then, we will find that with 10,000 iterations, such models are memory-hungry, creating huge files and slower refresh times. There are a number of dedicated software applications that will do all the hard work for us; some of these are directly compatible and interactive or even integrated with Microsoft Excel; some are not.

3.1.3 Key principle: Input variables with an uncertain future

All variables inherently have uncertain future values, otherwise we would call them Constants. We can describe the range of potential values by means of an appropriate probability distribution, of which there are scores, if not hundreds of different ones. We discussed a few of these in Volume II Chapter 4. We could (and we will) refer to this range of potential values as 'Uncertainty'. As is often the case, there is no universal source of truth about the use of the term 'Uncertainty' so let's define what we mean by it in this context.

Definition 3.1 Uncertainty

Uncertainty is an expression of the lack of sureness around a variable's eventual value, and is frequently quantified in terms of a range of potential values with an optimistic or lower end bound and a pessimistic or upper end bound

Here we have avoided using the definitive expressions of 'minimum' and 'maximum' in expressing a practical and reasonable range of uncertainty. In other words, we can often describe situations which are possible but which also fall outside the realm of reasonableness. However, for more pragmatic reasons when we input these uncertainty ranges into a Monte Carlo Simulation tool, we may often use statistical distributions which have absolute minima and maxima.

There is often a view expressed of either the 'Most Likely' (Mode) value or the 'Expected' value (Arithmetic Mean) within that range, although in the case of a Uniform Distribution a 'Most Likely' value is something of an oxymoron as all values are equally likely (*or unlikely if we have lots of them, and whether we are a glass half-full or half-empty personality*).

Each variable will have either a fixed number of discrete values it can take depending on circumstances, or an infinite number of values from a continuous range. If we were to pick an infinite number of values at random from a single distribution, we would get a representation of the distribution to which they belong. However, choosing an infinite number of things at random is not practical, so we'll have to stick with just having a large number of them instead!

For instance, if we tossed a coin a thousand times, or a thousand coins once each, then we would expect that around 500 would be 'Heads' and similarly around 500 would be 'Tails'. (*We'll discount the unlikely chance of the odd one or two balancing on its edge.*) This would reflect a 1 in 2 chance of getting either.

If we rolled a conventional die 6,000 times, we would expect approximately 1,000 of each of the faces numbered 1 to 6 to turn uppermost. This represents a discrete Uniform Distribution of the integers from 1 to 6.

If we divided the adult population of a major city into genders and ethnic groups we would expect that the heights of the people in each group to be approximately Normally Distributed.

So how can we exploit this property of large random numbers being representative of the whole? The answer is through Monte Carlo Simulation.

Let's consider a simple example of two conventional dice. Let's say we roll the two dice and add their values. In this case we can very simply compute the probability distribution for the range of potential values. Figure 3.1 summarises them. We have only one possible way of scoring 2 (1+1), and one way of scoring 12 (6+6), but we have six different ways of scoring 7:

$$1+6, 2+5, 3+4, 4+3, 5+2, 6+1$$

So, let's see what happens when we roll two dice at random 36 times. In a perfect world each combination would occur once and once only giving us the distribution above, but we just know that that is not going to happen, don't we? In fact we have tried this twice, with completely different results as shown in Figure 3.2. (*Not what we would call very convincing, is it?*) The line depicts the theoretical or true distribution derived in Figure 3.1; the histogram depicts our 36 random samples. The random selections look exactly that – random!

If we did it a third time and added all the random samples together so that we had 108 samples, we might get a slightly better result as we have in the left hand graph of Figure 3.3, but it's still not totally convincing. If we carried on and got 1,080 sample

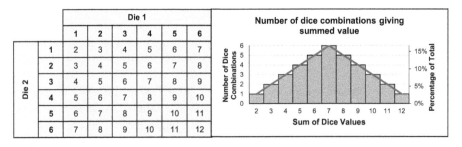

Figure 3.1 Probability Distribution for the Sum of the Values of Two Dice

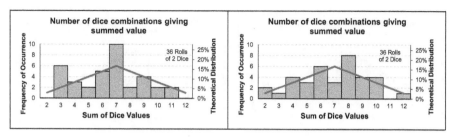

Figure 3.2 Sum of the Values of Two Dice Based on 36 Random Rolls (Twice)

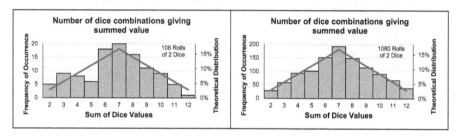

Figure 3.3 Sum of the Values of Two Dice Based on 108 and 1,080 Random Rolls

rolls of two dice, we would get something more akin to the right hand graph of Figure 3.3, which whilst it is not perfect fit, it is much more believable as evidence supporting the hypothesis that by taking a large number of random samples we will get a distribution that is more representative of the true distribution. So, let's continue that theme of increasing the number of iterations tenfold, and then repeat it all again just to make sure it was not a fluke result. We show both sets of results in Figure 3.4.

The two results are slightly different (for instance, look at the Summed Value of 6; in the left hand graph, the number of occurrence is slightly under 1,500 whereas it is slightly over that in the right hand graph.) However, we would probably agree that the two graphs are consistent with each other, and that they are both reasonably good representations of the true distribution, which is depicted by the line graph. In the grand scheme of estimating this difference is insignificant; it's another case of Accuracy being more important than Precision (Volume I Chapter 4). For completeness, we have shown the results of our two simulation runs of 10,800 random double dice rolls in Table 3.1, from which we can determine the cumulative probability or percentage frequency of occurrence. If we plot these in Figure 3.5, we can barely see the difference between the three lines. (*We might do if we have a magnifying glass handy.*)

What we have demonstrated here is the basic principles and procedure that underpin Monte Carlo Simulation:

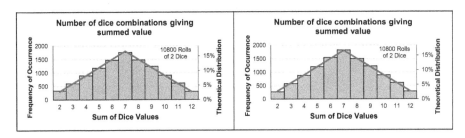

Figure 3.4 Sum of the Values of Two Dice Based on 10,800 Random Rolls

Table 3.1 Sum of the Values of Two Dice Based on 10,800 Random Rolls (Twice)

Sum of the Values of Two Dice	2	3	4	5	6	7	8	9	10	11	12	Total
Theoretical (True) Distribution	2.8%	8.3%	16.7%	27.8%	41.7%	58.3%	72.2%	83.3%	91.7%	97.2%	100%	
Simulation 1 Frequency	278	575	879	1235	1543	1823	1479	1203	892	600	293	10800
% of Total Number of Rolls	2.6%	5.3%	8.1%	11.4%	14.3%	16.9%	13.7%	11.1%	8.3%	5.6%	2.7%	
Cumulative % Frequency	2.6%	7.9%	16.0%	27.5%	41.8%	58.6%	72.3%	83.5%	91.7%	97.3%	100%	
Simulation 2 Frequency	332	621	920	1190	1473	1758	1439	1231	910	623	303	10800
% of Total Number of Rolls	3.1%	5.8%	8.5%	11.0%	13.6%	16.3%	13.3%	11.4%	8.4%	5.8%	2.8%	
Cumulative % Frequency	3.1%	8.8%	17.3%	28.4%	42.0%	58.3%	71.6%	83.0%	91.4%	97.2%	100%	

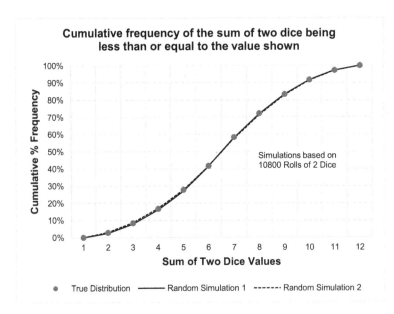

Figure 3.5 Cumulative Frequency of the Sum of Two Dice Being less than or Equal to the Value Shown

- Define the probability distribution of each variable in a model
- Select a value at random from each variable
- Compute the model value using these random choices
- Repeat the procedure multiple times
- Count the number of times each particular output total occurs, and convert it to a percentage of the total
- Stability in random sampling comes with an increased number of samples

Monte Carlo Simulation will model the likely output distribution for us.

3.1.4 Common pitfalls to avoid

One common mistake made by estimators and other number jugglers new to Monte Carlo Simulation is that they tend to use too small a sample size, i.e. too few iterations. In this case 'size does matter – bigger is better!' In any small sample size, it is always possible that we could get a freak or fluke result, the probability of which is very, very small (like winning the lottery that for some reason, has always eluded me!). Someone has to win it, but it is even more unlikely that someone will win the jackpot twice (although not impossible).

The stability that comes with large sample sizes was demonstrated in the last example where we had stability of the output when we used 10,800 iterations. (We used that number because it was a multiple of 36; there is no other hidden significance.)

Another common mistake people make is that they think that Monte Carlo Simulation is summing up complex probability distributions. This is quite understandable as the procedure is often described as taking a random number from a distribution AND another random number from another distribution AND … etc. Furthermore, the word 'AND' is normally associated mathematically with the additive operator PLUS. In the sense of Monte Carlo Simulations we should really say that AND should be interpreted as AND IN COMBINATION WITH, or DEPENDENT ON, the results of other things happening. In that sense we should describe it as the product of complex probability distributions, or perhaps liken it to Microsoft Excel's **SUMPRODUCT** function.

The net result is that Monte Carlo Simulation will always bunch higher probability values together around the Arithmetic Mean, narrowing the range of potential output values (recall our Measures of Central Tendency in Volume II Chapter 2), and 'dropping off' the very unlikely results of two or more extremely rare events occurring together.

Let's consider this in relation to 5 dice rolled together. The sum of the values could be as low as 5 or as high as 30, but these have only a 0.01286% chance each of occurring $(1/_6 \times 1/_6 \times 1/_6 \times 1/_6 \times 1/_6)$. In other words, on average we would only expect to get 5 1s or 5 6s, once in 10,000 tries. So, it's not actually impossible … just very improbable.

Without computing the probability of every possible sum of values between 5 and 30, Monte Carlo will show us that we can reasonably expect a value between 11 and 24 with

For the Formula-phobes: Monte Carlo Simulation narrows in on more likely values

It is a not uncommon mistake that people take the range of possible outcomes as being between the sum of the 'best case' values and the sum of the worst-case values. In an absolute sense, this is true, but the probability of getting either value at the bounds of the range is highly unlikely.

For instance, the chance of throwing a 6 on a die is 1 in 6; the chance of getting a double 6 is 1 in 36, or 1 in 6 multiplied by 1 in 6.

As we saw in Figure 3.1 each combination of the two dice have that same probability of $^1/_{36}$ of happening together, but there are six ways that we can score 7 meaning that the probability of scoring 7 is $^6/_{36}$ or $^1/_6$. So, we have to multiply the probabilities of each random event occurring and sum the number of instances that we get the same net result. The chance of two or more rare events occurring together is very small, we are more likely to get a mixture of events: some good, some bad. If we increase the number of dice, the number of physical combinations increases, and the chances of getting all very low or very high values in total get more unlikely, and gets more challenging to compute manually. However, we can simulate this with Monte Carlo without the need for all the mathematical or statistical number juggling. Remember …

Not all the good things in life happen together, and neither do all the bad things
… it might just feel that way sometimes.

90% Confidence. Here we have set up a simple model that generates a random integer between 1 and 6 (inclusive) using Microsoft Excel's **RANDBETWEEN(bottom, top)** function. We can then sum the 5 values to simulate the random roll of 5 dice. We can run this as many times as we like using a new row in Excel for every 'roll of the dice'. Using the function **COUNTIF(range, criteria)** we can then count the number of times we get every possible value between 5 and 30. We show an example in Figure 3.6 based on 10,000 simulation iterations.

For the Formula-phobes: 90% Confidence Interval

It might not be obvious from Figure 3.6 why the 90% Confidence Interval is between values 11 and 24. Unless we specify otherwise, the 90% Confidence Interval is taken to be the symmetrical interval between the 5% and 95% Confidence Levels. In our example:

- There are 283 values recorded with a score of 11 which accounts for a Confidence Level > 3.1% and ≤ 5.9%, implying that the 5% level must be 11
- There are 248 values recorded with a score of 24 which accounts for Confidence Level > 94.4% and ≤ 96.9%, implying that the 95% level must be 24

In fact, in this example the range 11 to 24 accounts for some 93.8% of all potential outcomes.

3.1.5 Is our Monte Carlo output normal?

If we look at the Probability Density Function for the range of possible outcomes, we will notice that it is very 'Normalesque' i.e. Bell-shaped (Figure 3.7) whereas when we were looking at only two dice (Figure 3.1), the output was distinctly triangular in shape.

This is no fluke; it is an early indication of phenomena described by the Central Limit Theorem and the Weak and Strong Laws of Large Numbers for independent identically distributed random samples. (*Don't worry, I'm not going to drag you through those particular proofs; there's plenty of serious textbooks and internet sources that will take you through those particular delights.*)

However, although this tendency towards a Normal Distribution is quite normal, it is not always the result!

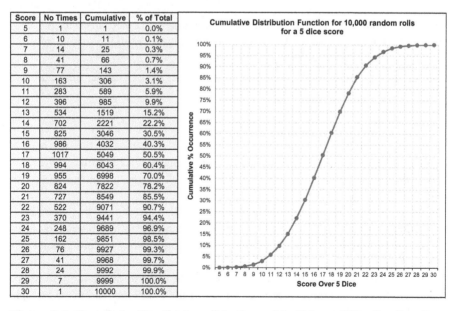

Score	No Times	Cumulative	% of Total
5	1	1	0.0%
6	10	11	0.1%
7	14	25	0.3%
8	41	66	0.7%
9	77	143	1.4%
10	163	306	3.1%
11	283	589	5.9%
12	396	985	9.9%
13	534	1519	15.2%
14	702	2221	22.2%
15	825	3046	30.5%
16	986	4032	40.3%
17	1017	5049	50.5%
18	994	6043	60.4%
19	955	6998	70.0%
20	824	7822	78.2%
21	727	8549	85.5%
22	522	9071	90.7%
23	370	9441	94.4%
24	248	9689	96.9%
25	162	9851	98.5%
26	76	9927	99.3%
27	41	9968	99.7%
28	24	9992	99.9%
29	7	9999	100.0%
30	1	10000	100.0%

Figure 3.6 Cumulative Distribution of the Sum of the Values of Five dice Based on 10,000 Random Rolls

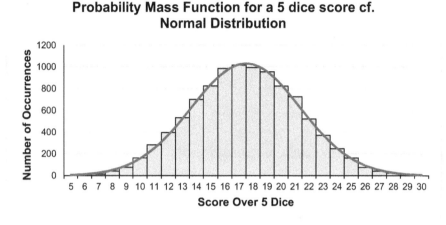

Figure 3.7 Probability Mass Function for a 5 Dice Score cf. Normal Distribution

---

Caveat augur

Whilst modelling uncertainty in a system of multiple variables can often appear to be Normally Distributed, do not make the mistake of assuming that it always will be. It depends on what we are modelling.

For instance, let's look at a case where the output is not Normal. Consider a continuous variable such as the height of school children. Now we know that the heights of adult males are Normally Distributed, as are the heights of adult females. The same can be inferred also for children of any given age. To all intents and purposes, we can identify three variables:

- **Gender** – which for reasons of simplicity we will assume to be a discrete uniform distribution, i.e. 50:50 chance of being male or female (*rather than a slightly less imprecise split of 49:51*)
- **Age** – which we will also assume to be a discrete uniform distribution. In reality this would be a continuous variable but again for pragmatic reasons of example simplicity we will round these to whole years (albeit children of a younger age are

more likely to be precise and say they are $6^7/_{12}$ or whatever). We are assuming that class sizes are approximately the same within a school

- **Height for a given Gender and Age** – we will assume to be Normally Distributed as per the above

Now let's visit a virtual primary school and sample the heights of all its virtual pupils. No, let's visit all the virtual primary schools in our virtual city and see what distribution we should assume for the height of a school child selected at random.

> The child selected could be a Girl or a Boy, and be any age between 5 years and 11 (assuming the normal convention of rounding ages down to the last birthday).
>
> Many people's first intuitive reaction is that it would be a Normal Distribution, which is quite understandable, but unfortunately it is also quite wrong. (*If you were one of those take heart in the fact that you are part of a significant majority.*)
>
> When they think about it they usually realise that at the lower age range their heights will be Normally Distributed; same too at the upper age range, and ... everywhere between the two. Maybe it's a flat-topped distribution with Queen Anne Legs at the leading and trailing edges (i.e. slightly S-Curved)?

However, although that is a lot closer to reality, it is still not quite right. Figure 3.8 illustrates an actual result from such random sampling ... the table top is rounded and sloping up from left to right, more like a tortoise. For the cynics amongst us (*and what estimator isn't endowed with that virtue?*) who think perhaps that we have a biased sample of ages, Figure 3.9 shows that the selection by age is Uniformly Distributed and as such is consistent with being randomly selected.

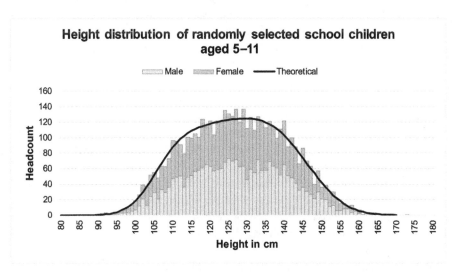

Figure 3.8 Height Distribution of Randomly Selected School Children Aged 5–11

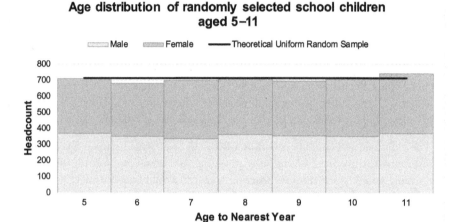

Figure 3.9 Age Distribution of Randomly Selected School Children Aged 5–11

Let's analyse what is happening here. There is little significant difference between the heights of girls and boys at primary school age. Furthermore, the standard deviation of their height distributions at each nominal age of 5 through to 11 is also fairly consistent at around 5.5% to 5.6%. (Standard Deviation based on the spread of heights between the Mean Nominal Age and the 2nd Percentile six months younger and the 98th Percentile six months older). These give us the percentile height growth charts shown in Figure 3.10, based on those published by Royal College of Paediatrics and Child Health (RCPCHa,b, 2012). (It is noted that there are gender differences but these are relatively insignificant for this age range … *perhaps that's a hint of where this discussion is going.*)

We have calculated the theoretical distribution line in Figure 3.8 using the sum of a Normal Distribution for each nominal age group for girls and boys using the 50th Percentile (50% Confidence Level) from Figure 3.10. The 96% Confidence Interval between 2% and 98% has been assumed to be 4 Standard Deviations (offset by 6 months on either side). Figure 3.11 shows the individual Normal Distributions which then summate to the overall theoretical distribution (assuming the same number of boys and girls in each and every age group). We will have noticed that the age distributions get wider and shorter … but their areas remain the same; this is due to the standard deviation increasing in an absolute sense as it is a relatively constant percentage of the mean. The net result is that there is an increased probability of children of any given height, and the characteristic, albeit slight sloping table top.

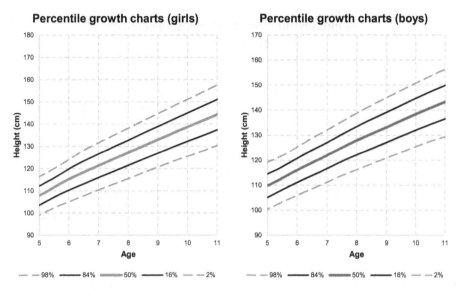

Figure 3.10 Percentile Height School Children Ages 5–11
Source: Copyright © 2009 Royal College of Paediatrics and Child Health

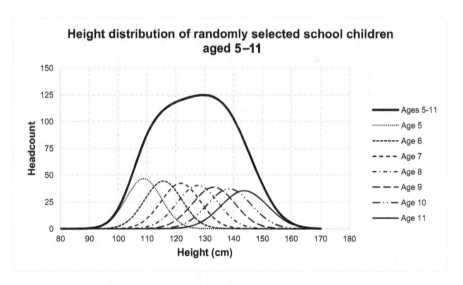

Figure 3.11 Height Distribution of Randomly Selected School Children Aged 5–11

Now let's extend the analysis to the end of secondary school and 6th form college, nominally aged 18. Let's look at all girls first in the age range from 5 to 18 (Figure 3.12). Following on from the primary years we get a short plateauing on

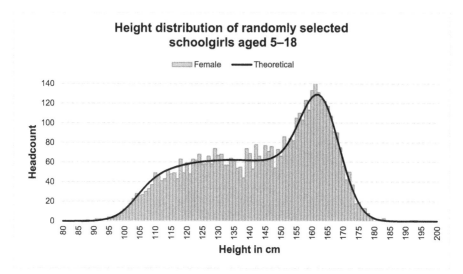

Figure 3.12 Height Distribution of Randomly Selected Schoolgirls Aged 5–18

the number of schoolgirls over an increasing height, before it rises dramatically to a peak around 163.5 cm.

Figure 3.13 explains why this happens. After age 13, girls' growth rate slows down and effectively reaches a peak between the ages 16 and 18. This gives us more girls at the taller height range. Figure 3.14 shows the primary and secondary age groups separately just for clarity. Whereas primary school aged children were more of a rounded mound with a slight slope, the secondary school aged girls are 'more normal' with a small negative skew.

Now let's turn our attention to the boys through primary and secondary school including 6th form college (aged 5–18). It's time for an honesty session now… how many of us expected it to be the Bactrian Camel Distribution shown in Figure 3.15?

No, the dip is nothing to do with teenage boys having a tendency to slouch. Figure 3.16 provides us with the answer; it is in the anomalous growth of boys through puberty, where we see a slowdown in growth after age 11 (in comparison to girls – right hand graph) before a surge in growth in the middle teen years, finally achieving full height maturity at 18 or later.

As a consequence of this growth pattern we get:

- Less boys in the 140–160 cm height range as they accelerate through this range
- The slowdown immediately prior to this growth surge causes an increased number of boys in the height range 135–140 cm in comparison to girls.
- After age 16, boys' growth rate slows again, giving a more Normal Distribution around 175 cm as they reach full expected growth potential

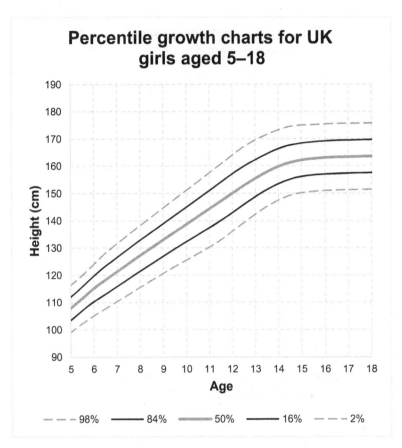

Figure 3.13 Percentile Growth Charts for UK Girls Aged 5–18
Source: Copyright © 2009 Royal College of Paediatrics and Child Health

- For completeness, we have shown primary and secondary school aged boys separately in Figure 3.17

Overall across all schoolchildren aged 5.18, we get the distribution in Figure 3.18. Collectively from this we might conclude the following:

- Overall randomly selected children's height is not Normally Distributed
- Teenage boys are Statistical Anomalies, *but the parents amongst us will know that already and be nodding wisely*
- Collectively, girls become Normal before boys! (*My wife would concur with that, but I detect an element of bias in her thinking*)

    More importantly …

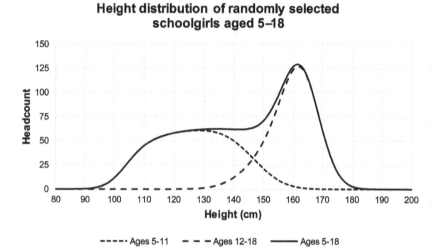

Figure 3.14 Height Distribution of Randomly Selected Schoolgirls Aged 5–18

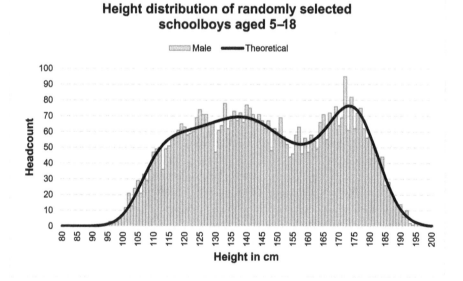

Figure 3.15 Height Distribution of Randomly Selected Schoolboys Aged 5–18

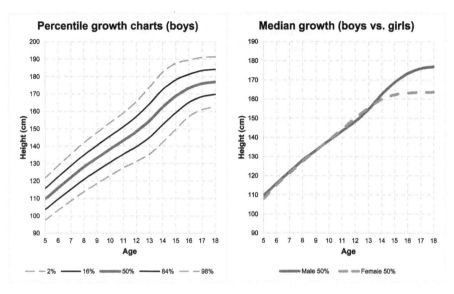

Figure 3.16 Percentile Growth Charts for UK Boys Aged 5–18
Source: Copyright © 2009 Royal College of Paediatrics and Child Health

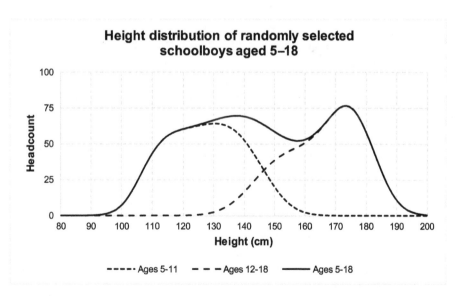

Figure 3.17 Height Distribution of Randomly Selected Schoolboys Aged 5–18

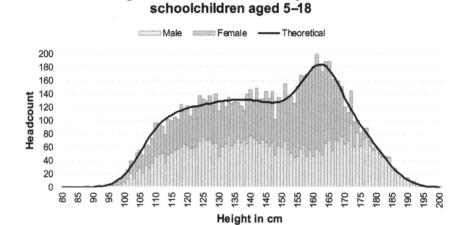

Figure 3.18 Height Distribution of Randomly Selected Schoolchildren Aged 5–18

- Monte Carlo Simulation is a very useful tool to understand the nature of the impact of multiple variables, and that the output is not necessarily Normally Distributed, or even one that we might intuitively expect
- Don't guess the output; it's surprising how wrong we can be

3.1.6 Monte Carlo Simulation: A model of accurate imprecision

When we are building a Monte Carlo Simulation Model, there is **no need** to specify input parameters to a high degree of precision, other than where it provides that all-important TRACEability link to the Basis of Estimate.

Input variables can be in any measurement scale or currency, but at some point, there has to be a conversion in the model to allow an output to be expressed in a single currency. In this scenario, the term currency is not restricted to financial or monetary values, but could be physical or time-based values:

- These conversion factors could be fixed scale factors (such as Imperial to metric dimensions) in which case there is no variability to consider
- The conversion factors could themselves be variables with associated probability distributions such as Currency Exchange Rates or Escalation Rates

---

Caveat augur

Don't waste time and energy or spend sleepless nights on unnecessary imprecision. Take account of the range of the input variables' scales in relation to the whole model.

In terms of input parameters for the input variable distributions, such as the Minimum (or Optimistic), Maximum (or Pessimistic) or Mode (Most Likely) values, we do not need to use precise values; accurate rounded values are good enough.

For example, the cumulative probability of getting a value that is less than or equal to £ 1,234.56789 k in a sample distribution cannot be distinguished from a statistical significance point of view from getting the value £ 1,235 k.

Using the rounded value as an input will not change the validity of the output.

However, resist the temptation to opt for the easy life of a plus or minus a fixed percentage of the Most Likely (see the next section) unless there are compelling reasons that can be captured and justified in the Basis of Estimate; remember the principles of TRACEability (Transparent, Repeatable, Appropriate, Credible and Experientially-based). Remember that input data often has a natural skew.

For an example of this 'accurate imprecision' let's turn our attention to one of the most common, perhaps popular, uses of Monte Carlo Simulation in the world of estimating and scheduling ... modelling uncertainty across a range of input values. In Table 3.2 we have shown ten cost variables each with its own defined distribution. For reference, we have summated the Minimum and Maximum values across the variables, along with the sum of their modes and means.

If we run the model with 10,000 iterations, we will get an output very similar to that shown in Figures 3.19 and 3.20. Every time we run the model, even with this number of iterations, we will get precisely different but very consistent results. In other words, Monte Carlo Simulation is a technique to use to get accurate results with acceptable levels of imprecision. For instance, if we were to re-run the model, the basic shape of the Probability Density Function (PDF) graph in Figure 3.19 will remain the same, but all the individual mini-spikes will change (as shown in the upper half of Figure 3.21).

In contrast, the Cumulative Distribution Function (CDF) S-Curve of the probability of the true outcome being less than or equal to the value depicted will hardly change at all if we re-run the model, as illustrated in the lower half of Figure 3.21. (*We're talking of 'thickness of a line' differences only here.*)

Table 3.3 emphasises the consistency by comparing some of the key output statistics from these two simulations using common input data.

From the common input data in our earlier Table 3.2, the output results in Table 3.3 and either result in Figure 3.21 we can observe the following:

Table 3.2 Example Monte Carlo Simulation of Ten Independently Distributed Cost Variables

Cost Element	Distribution Type	Potential Cost Range £k (Input Data)			Mean	Median
		Minimum	Mode	Maximum		
Cost Item 1	Uniform	50	80	110	80	80
Cost Item 2	PERT-Beta	80	101	150	105.67	104.57
Cost Item 3	Beta	60	80	120	82.5	81.85
Cost Item 4	Triangular	20	50	80	50	50
Cost Item 5	Triangular	30	50	100	60	56.46
Cost Item 6	Beta	75	100	150	105	103.93
Cost Item 7	PERT-Beta	90	100	120	101.67	101.27
Cost Item 8	Beta	70	80	110	82.5	81.85
Cost Item 9	Uniform	35	45	55	45	45
Cost Item 10	Triangular	90	95	105	96.67	96.12
Total		600	781	1100	809.00	809

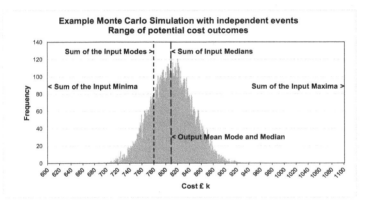

Figure 3.19 Sample Monte Carlo Simulation PDF Output Based on 10,000 Iterations

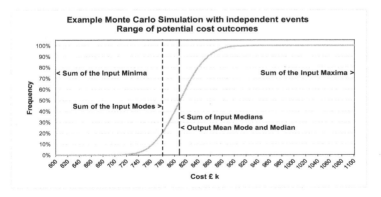

Figure 3.20 Sample Monte Carlo Simulation CDF Output Based on 10,000 Iterations

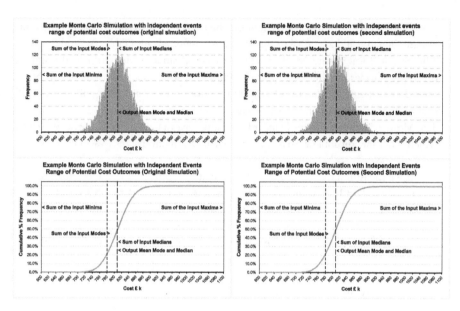

Figure 3.21 Comparison of Two Monte Carlo Simulations of the Same Cost Model

Table 3.3 Comparison of Summary Output Statistics Between Two Corresponding Simulations

	Original Simulation		Second Simulation		Observation
	Value £k	Confidence Level	Value £k	Confidence Level	
Minimum	686	0.01%	689	0.02%	Precise Minimum is often more volatile
Median	809	50.00%	809	50.00%	**Median is relatively stable**
Mode	809	48.84%	811	51.17%	Precise Mode is often more volatile
Mean, μ	809.6	50.74%	809.321	50.45%	**Mean is relatively stable**
Maximum	941	100.00%	942	100.00%	Precise Maximum is often more volatile
Left Range	123		122		Reflects changes in Precise Maximum and Mode
Right Range	132		131		Reflects changes in Precise Minimum and Mode
Total Range	255		253		Reflects changes in Precise Minimum and Maximum
Std Dev, σ	35.7		35.8		**Standard Deviation is relatively stable**
μ - 3σ	702.6	0.06%	701.8	0.05%	
μ + 3σ	916.6	99.86%	916.8	99.87%	
6σ Range	214.1	99.81%	215.0	99.82%	Normal Distribution would be 99.73%
μ - 2σ	738.2	2.11%	737.7	2.02%	
μ + 2σ	881.0	97.78%	881.0	97.79%	
4σ Range	142.7	95.67%	143.3	95.67%	Normal Distribution would be 95.45%
Skew	0.050		0.064		Normal Distribution would be zero
Kurtosis	-0.116		-0.177		Normal Distribution would be zero

- With the exception of the two variables with Uniform Distributions, all the other input variables have positively skewed distributions (i.e. the Mode is closer to the Minimum than the Maximum (Table 3.2)
- The output distribution of potential cost outcomes is almost symmetrical as suggested by the relatively low absolute values of the Skewness statistic (Table 3.3)
- The output distribution is significantly narrower than that we would get by simply summing the minima and the maxima of the input distributions (Figure 3.21)

- Consistent with a symmetrical distribution, the rounded Mean and Median of the Output are the same and are equal to the sum of the input Means (Tables 3.2 and 3.3)
- The precise 'local output Mode' is closer to the Mean and Median than the sum of the input Modes, further suggesting a symmetrical distribution (Table 3.3). Note: on a different simulation run of 10,000 iterations, the local mode will always drift slightly in value due to changes in chance occurrences
- The Mean +/- three times the Standard Deviation gives a Confidence Interval of 99.84% which is not incompatible with that of a Normal Distribution at 99.73% (Table 3.3)

  o   If we measured the Confidence Interval at +/- one Standard Deviation or +/- two Standard Deviations around the Mean we would get 67.59% and 95.77% respectively. This also compares favourably with the Normal Distribution intervals of 68.27% and 95.45%

  o   The Excess Kurtosis being reasonably close to zero is also consistent with that of an approximately Normal Distribution

However, if only life was as simple as this …

Caveat augur

This model assumes that all the input variable costs are independent of each other, and whilst this may seem a reasonable assumption, it is often not true in the absolute sense. For example, overrunning design activities may cause manufacturing activities to overrun also, but manufacturing as a function is quite capable of overrunning without any help from design!

Within any system of variables there is likely to be a degree of loose dependence between ostensibly independent variables. They share common system objectives and drivers.

We will return to the topic of partially correlated input variables in Section 3.2.

3.1.7 What if we don't know what the true Input Distribution Functions are?

At some stage we will have to make a decision about the choice of input distribution. This is where Monte Carlo Simulation can be forgiving to some extent. The tendency of models to converge towards the central range of values will happen regardless of the distributions used so long as we get the basic shape correct, i.e. we don't want to use negatively skewed distributions when they should be positively skewed, but even then, Monte Carlo has the 'magnanimity' to be somewhat forgiving.

Rule of Thumb

However, as a Rule of Thumb, distributions of independent variables of cost or time are more likely to be positively rather than negatively skewed. We cannot rule out symmetrical distributions if the input is at a sufficiently high level in terms of systems integration. If we were to model the basic activities that result in the higher level integrated system then, as we have already seen, they would naturally converge to a symmetrical distribution that can be approximated to a Normal Distribution. Neither can we rule out negatively skewed distributions of other variables that might be constituent elements of a cost or schedule calculation such as the performance we might expect to achieve against some standard, norm or benchmark, but for basic cost and schedule, we should expect these to be positively skewed.

For the Formula-phobes: Why are cost and time likely to be positively skewed?

Suppose we have to cross the precinct in town to get to the Bank, a distance of some 100m.

We might assume that at a reasonable average walking pace of around four miles per hour, it will take us around a minute. Let's call that our Most Likely Time.

The best we could expect to achieve if we were world class athletes, would be around ten seconds, giving us what we might consider to be an absolute minimum. (*So, that's not going to happen in my case, before anyone else says it.*)

The worst time we could reasonably expect, could be considerably more than our Most Likely Time. The town centre may be very busy and we may have to weave our way around other people, and we're in no rush, we can take as long as we like, do a bit of window shopping on route ... and then there's the age factor for some of us, and ... in these shoes?!

The net result is that our capacity to take longer is greater than our capacity to do it more quickly. We have an absolute bound at the lower end (*it must be greater than zero*), but it's relatively unbounded at the top end (*life expectancy permitting!*).

Caveat augur

The natural tendency for positive skewness in input variables of time and cost is a valid consideration at the lowest detail level. If we are using higher level system cost or time summaries, these are more likely to be Normally Distributed ... as we are about to discover.

If we were to include these remote, more extreme values in our Monte Carlo analysis, they are unlikely to occur by random selection, and if one did, it would be a bit of a 'one-off' that we would ignore in relation to the range of realistic values we could reasonably expect. That though may lead us to wonder whether it really matters what distributions we input to our Monte Carlo Model. Let's consider now what might those same basic distribution shapes be that we might 'get away with' substituting (summarised in Table 3.4), and see how they perform. Note that we can mirror the positively skewed distributions in this table to give us their negatively skewed equivalents should the eventuality arise.

Note: If we do know the likely distribution because we have actual data that has been calibrated to a particular situation, then we should use it; why go for the rough and ready option? Remember the spirit of TRACEability (Transparent, Repeatable, Appropriate, Credible and Evidence-based).

More often though, we won't have the luxury of a calibrated distribution and we will have to rely on informed judgement, and that might mean having to take a sanity check.

Table 3.4 Example Distributions that Might be Substituted by Other Distributions of a Similar Shape

True Distribution	Potential Substitute Distribution	Comments
+ve Skewed Beta Distribution ($\beta > \alpha > 2$)	+ve Skewed Triangular Distribution	We need to truncate the Maximum Value to avoid the long trailing leg of the Beta Distribution where the probability of getting a value is very small
+ve Skewed Peaky Beta Distribution ($\beta \geq \alpha > 8$)	Normal Distribution	With peaky Beta Distributions, the probability concentrates around the Mode to such an extent that it can be approximated by a Normal Distribution around the Mode (offset to the left of the Beta range)
Normal Distribution or Symmetrical Beta Distributions	Symmetrical Triangular Distribution	The Triangular Minimum and Maximum are assumed to be equivalent to 2.33 Standard Deviations either side of the Normal Distribution Mean (see Volume II Chapter 4)

For instance, if we are considering using a Beta Distribution but we don't know the precise parameter values (i.e. it has not been calibrated to actual data) then we might like to consider plotting our distribution first using our 'best guess' at the parameter values (i.e. alpha, beta, AND the Start and End points.) If the realistic range of the distribution does not match our intentions or expectations then we can modify our input parameters, or use a more simplistic Triangular Distribution.

If you are concerned that we may be missing valid but extreme values by using a Triangular Distribution, then we can always consider modelling these as Risks (see Section 3.3.1) but as we will see it may not be the best use of our time.

Before we look at the effects on making substitutions, let's examine the thought processes and see what adjustments we might need to make in our thinking.

- If the only things of which we are confident are the lower and upper bounds, i.e. that the Optimistic and Pessimistic values cover the range of values that are likely to occur in reality, then we should use a Uniform Distribution because we are saying in effect that no single value has a higher chance of arising than any other … to the best of our knowledge.
- If we are comfortable with our Optimistic value and that the Pessimistic value covers what might be reasonably expected, but excludes the extreme values or doomsday scenarios covered by the long trailing leg of a Beta Distribution, then provided that we are also confident that one value is more likely than any other, then we can be reasonably safe in using a Triangular Distribution
- If we are reasonably confident in our Optimistic and Most Likely values but are also sure that our Pessimistic value is really an extreme or true 'worst case' scenario, then we should either use a Beta Distribution … or truncate the trailing leg and use a Triangular Distribution with a more reasonable pessimistic upper bound. Table 3.5 illustrate how close we can approximate a range of positively and negatively skewed Beta Distributions by Triangular Distributions

From a practical point of view, if we know the Beta Distribution shape, (and therefore its parameters) then why would we not use them anyway? There is no need to approximate them just for the sake of it (*who needs extra work?*). However, suppose we believe that we have a reasonable view of the realistic lower and upper bounds but that we want to make sure that we have included the more extreme values in either the leading or trailing legs of a wider Beta Distribution, how can we use our Triangular approximation to get the equivalent Beta Distribution?

However, whilst we can create Triangular Distributions which have the same key properties as Beta Distributions such that the Modes and Means match, the relationship is too intertwined and complex to be of any practical use, as illustrated by the Formulaphile call-out. We could of course find the best fit solution using Microsoft Excel's Solver and performing a Least Squares algorithm but that again begs the question of why we are looking at an approximation if we already know what we think the true distribution is? We are far better in many instances asking ourselves what the reasonable lower and upper bounds of a distribution are and using those with a Triangular Distribution.

Table 3.5 Beta and Triangular Distributions with Similar Shapes

For the Formula-philes: Substituting the Beta Distribution with a Triangular Distribution

Consider a Triangular Distribution with a start point B and an end point E and a mode,. Consider also a Beta Distribution with parameters α and β, a start point S and finish point F such that its mode, $\hat{M}$, coincides with that of the Triangular Distribution. Let the two distributions have a common mean, $\bar{M}$ also.

From the standard properties of a Beta Distribution, the Mode $\hat{M}$ occurs at:

$$\hat{M} = S + \frac{(\alpha - 1)}{(\alpha + \beta - 2)}(F - S) \quad (1)$$

... and the Mean $\bar{M}$ of the Beta Distribution occurs at:

$$\bar{M} = S + \frac{\alpha}{(\alpha + \beta)}(F - S) \quad (2)$$

From the standard properties of a Triangular Distribution the Mean $\bar{M}$ occurs at:

$$\bar{M} = \frac{B + \hat{M} + E}{3} \quad (3)$$

From (2) and (3):
Substituting (1) in (4)

$$\frac{B + \hat{M} + E}{3} = S + \frac{\alpha}{(\alpha + \beta)}(F - S) \quad (4)$$

and simplifying: $$B + S + \frac{(\alpha - 1)}{(\alpha + \beta - 2)}(F - S) + E = 3S + \frac{3\alpha}{(\alpha + \beta)}(F - S) \quad (5)$$

Re-arranging (5) and simplifying: $$\left(\frac{3\alpha}{(\alpha + \beta)} - \frac{(\alpha - 1)}{(\alpha + \beta - 2)} \right)(F - S) = B + E - 2S \quad (6)$$

Expanding and simplifying (6): $$\left(\frac{2\alpha^2 + 2\alpha\beta - 5\alpha + \beta}{(\alpha + \beta)(\alpha + \beta - 2)} \right)(F - S) = B + E - 2S \quad (7)$$

Factorising (7): $$\left(\frac{(2\alpha + 1)(\alpha + \beta - 3) + 3}{(\alpha + \beta)(\alpha + \beta - 2)} \right)(F - S) = B + E - 2S \quad (8)$$

Simplifying (8): $$F - S = \frac{(B + E - 2S)(\alpha + \beta)(\alpha + \beta - 2)}{(2\alpha + 1)(\alpha + \beta - 3) + 3} \quad (9)$$

… which expresses the range of the 'equivalent' Beta Distribution in relation to a Triangular Distribution with the same Mean and Mode. We just need to know the sum of the two Beta Distribution parameters, and to have an idea of the Beta Distribution Start Point if it is Positively Skewed.

… not really the snappiest or most practical of conversion formulae, is it?

I can see that some of us are not convinced. Let's see what happens when we start switching distributions. Using the data from our previous Table 3.2 and plotting the five Beta Distributions, we might accept that the Triangular Distribution in Figure 3.22 are not unreasonable approximations.

Now let's see what happens if we make these Triangular substitutions from Figure 3.22 in our Model Carlo Model. Figure 3.23 illustrates the result on the right hand side in comparison with the original data from Figure 3.19 reproduced on the left. Now let's play a game of '*Spot the Difference*' between the left and right hand graphs. Apart from the exact positions of the individual 'stalagmites' in the upper graph, it is hard to tell any real difference in the lower cumulative graph.

Table 3.6 summarises and compares some of the key statistics. Note that in this instance we are using the same set of random numbers to generate the Monte Carlo Simulations as we did in the original simulation; they're just 'driving' different distributions. (*Just in case you thought it might make a difference by fluke chance.*)

Now let's try something inappropriate (*no, not in that way, behave*); let's forget about making the adjustment to truncate the Beta Distribution Maximum and simply replace it with a Triangular Distribution; we've done just that in Figure 3.24 and Table 3.7. This time there is a marked difference with the central and right areas of the output

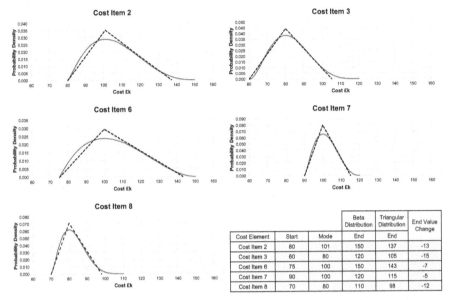

Cost Element	Start	Mode	Beta Distribution End	Triangular Distribution End	End Value Change
Cost Item 2	80	101	150	137	-13
Cost Item 3	60	80	120	105	-15
Cost Item 6	75	100	150	143	-7
Cost Item 7	90	100	120	115	-5
Cost Item 8	70	80	110	98	-12

Figure 3.22 Approximation of 5 Beta Distributions by 5 Triangular Distributions

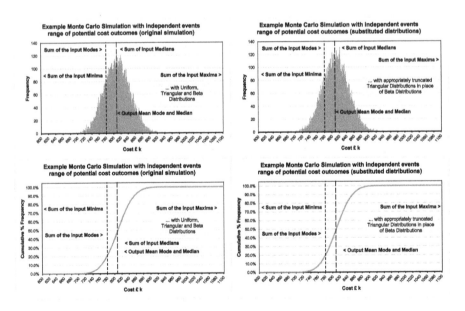

Figure 3.23 Comparison of Two Monte Carlo Simulations with Appropriate Substituted Distributions

Table 3.6 Comparison of Two Monte Carlo Simulations with Appropriate Substituted Distributions

	Original Simulation		Substitute Simulation		
	Value £k	Confidence Level	Value £k	Confidence Level	Observation
Minimum	686	0.01%	669	0.00%	Reflects increased chance of lower values
Median	809	50.00%	825	50.00%	Average of Minimum and Maximum
Mode	809	48.84%	818	58.84%	Precise Mode is always more volatile
Mean, μ	809.6	50.74%	824.8	67.19%	Average of Minimum and Maximum
Maximum	941	100.00%	981	100.00%	Reflects increased chance of higher values
Left Range	123		149		Reflects the wider range of values
Right Range	132		163		
Total Range	255		312		
Std Dev, σ	35.7		44.9		
μ - 3σ	702.6	0.06%	690.1	0.02%	
μ + 3σ	916.6	99.86%	959.5	100.00%	
6σ Range	214.1	99.81%	269.4	99.98%	Normal Distribution would be 99.73%
μ - 2σ	738.2	2.11%	735.0	1.56%	
μ + 2σ	881.0	97.78%	914.6	99.81%	
4σ Range	142.7	95.67%	179.6	98.25%	Normal Distribution would be 95.45%
Skew	0.050		-0.025		Normal Distribution would be zero
Kurtosis	-0.116		-0.153		Normal Distribution would be zero

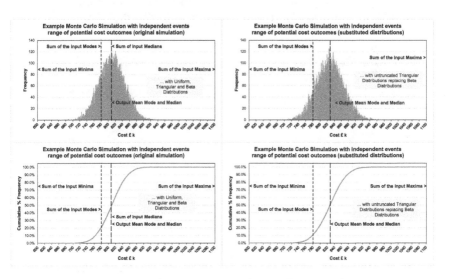

Figure 3.24 Comparison of Two Monte Carlo Simulations with Inappropriate Substituted Distributions (1)

distribution moving more to the right; the left hand 'start' area is largely unchanged. Overall, however, the basic output shape is still largely 'Normalesque' with its symmetrical bell-shape, leading us to re-affirm perhaps that system level costs are likely to be Normally Distributed rather than positively skewed.

However, if we had simply replaced all the distributions with Uniform Distributions on the grounds that we weren't confident in the Most Likely value, we would have widened the entire output distribution even more, but on both sides, but yet it would have still remained basically Normalesque in shape, as illustrated in Figure 3.25 and Table 3.8.

Table 3.7 Comparison of Two Monte Carlo Simulations with Inappropriate Substituted Distributions (1)

	Original Simulation		Substitute Simulation		
	Value £k	Confidence Level	Value £k	Confidence Level	Observation
Minimum	686	0.01%	694	0.02%	Precise Minimum is often more volatile
Median	809	50.00%	827	50.00%	Shift to the right reflects the increased probability of larger values with a Triangular Distribution instead of a Beta Distribution with the same range
Mode	809	48.84%	824	65.52%	
Mean, μ	809.6	50.74%	827.6	69.82%	
Maximum	941	100.00%	968	100.00%	
Left Range	123		130		Reflects shift in Mode to the right
Right Range	132		144		Reflects the increased chance of higher values
Total Range	255		274		Reflects the wider range of values
Std Dev, σ	35.7		38.0		
μ - 3σ	702.6	0.06%	713.7	0.22%	
μ + 3σ	916.6	99.86%	941.5	100.00%	
6σ Range	214.1	99.81%	227.8	99.78%	Normal Distribution would be 99.73%
μ - 2σ	738.2	2.11%	751.7	5.16%	
μ + 2σ	881.0	97.78%	903.5	99.57%	
4σ Range	142.7	95.67%	151.8	94.41%	Normal Distribution would be 95.45%
Skew	0.050		0.058		Normal Distribution would be zero
Kurtosis	-0.116		-0.096		Normal Distribution would be zero

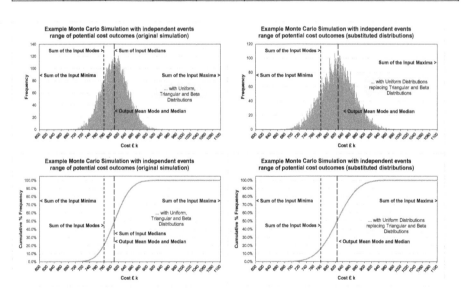

Figure 3.25 Comparison of Two Monte Carlo Simulations with Inappropriate Substituted Distributions (2)

Table 3.8 Comparison of Two Monte Carlo Simulations with Inappropriate Substituted Distributions (2)

	Original Simulation		Substitute Simulation		Observation
	Value £k	Confidence Level	Value £k	Confidence Level	
Minimum	686	0.01%	669	0.00%	Reflects increased chance of lower values
Median	809	50.00%	825	50.00%	Average of Minimum and Maximum
Mode	809	48.84%	818	58.84%	Precise Mode is always more volatile
Mean, μ	809.6	50.74%	824.8	67.19%	Average of Minimum and Maximum
Maximum	941	100.00%	981	100.00%	Reflects increased chance of higher values
Left Range	123		149		Reflects the wider range of values
Right Range	132		163		
Total Range	255		312		
Std Dev, σ	35.7		44.9		
μ - 3σ	702.6	0.06%	690.1	0.02%	
μ + 3σ	916.6	99.86%	959.5	100.00%	
6σ Range	214.1	99.81%	269.4	99.98%	Normal Distribution would be 99.73%
μ - 2σ	738.2	2.11%	735.0	1.56%	
μ + 2σ	881.0	97.78%	914.6	99.81%	
4σ Range	142.7	95.67%	179.6	98.25%	Normal Distribution would be 95.45%
Skew	0.050		-0.025		Normal Distribution would be zero
Kurtosis	-0.116		-0.153		Normal Distribution would be zero

We can conclude from this that there is no way of telling from the output distribution whether we have chosen the right input variable's distributions of uncertainty as it will always be generally Normalesque, i.e. bell-shaped and symmetrical, where we are aggregating costs or time. (*This may not be the case when we use Monte Carlo in other ways, for instance when we consider risk and opportunity ... we'll be coming to those soon.*)

In order to make an appropriate choice of distribution where we don't know the true distribution then we should try to keep it simple, and ask ourselves the checklist questions in Table 3.9. For those of us who prefer diagrams, these are shown in Figures 3.26 and 3.27 for basic input levels, and system level input variables.

In terms of our 'best guess' at Beta Distribution shape parameters (alpha and beta), we can choose them based on the ratio of the Mode's relative position in the range between the Optimistic and Pessimistic values. (The alpha and beta parameters will always be greater than one.) Typical values are shown in Table 3.10, but as Figure 3.28 illustrates, several Beta Distributions can be fit to pass through the same Start, Mode and Finish points. However, the peakier the distribution is (i.e. greater values of parameters alpha and beta combined), then the more likely it is that the minimum and maximum values will be extreme values and therefore may not be representative of the Optimistic and Pessimistic values intended.

Table 3.9 Making an Appropriate Informed Choice of Distribution for Monte Carlo Simulation

Is the input variable at the system level, i.e. the net result of multiple lower level variables?	Do we have a view on the optimistic and pessimistic bounds?	Are the optimistic and pessimistic bounds symmetrical around the Most Likely value?	Do we have a view of a single Most Likely value?	Are the optimistic and pessimistic bounds the absolute best and worst case values?	Distribution to Consider
Either	No	N/A	No	N/A	*Nothing to Simulate*
Either	No	N/A	Yes	N/A	*Nothing to Simulate*
No	Yes	Either	Yes	No	Triangular
No	Yes	Either	Yes	Yes	Beta (See Note 1)
No	Yes	N/A	No	Either	Uniform
Yes	Yes	N/A	No	No	Normal (See Notes 2, 3)
Yes	Yes	N/A	No	Yes	Normal (See Note 2, 4)
Yes	Yes	Yes	Yes	No	Normal (See Note 3)
Yes	Yes	Yes	Yes	Yes	Normal (See Note 4)
Yes	Yes	No	Yes	No	Triangular
Yes	Yes	No	Yes	Yes	Beta (See Note 1)

Notes: 1. Choose parameters such that (Mode − Start) / (End − Start) = $(\alpha - 1) / (\alpha + \beta - 2)$

2. Take the Mean as the average of the optimistic and pessimistic values

3. Take the Standard Deviation as the optimistic-pessimistic range divided by 4

4. Take the Standard Deviation as the optimistic-pessimistic range divided by 6

Caveat augur

These checklists decision and flow diagrams are aimed at modelling cost and schedule estimates and should not be construed as being applicable to all estimating and modelling scenarios, where other specific distributions may be more appropriate.

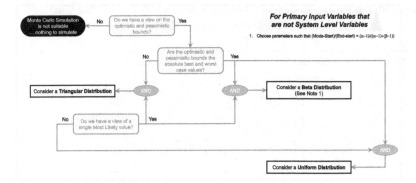

Figure 3.26 Making an appropriate informed choice of Distribution for Monte Carlo Simulation (1)

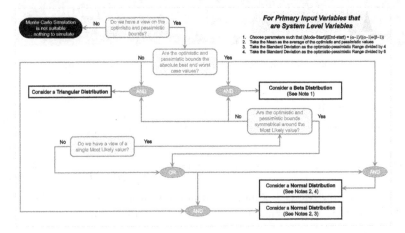

Figure 3.27 Making an Appropriate Informed Choice of Distribution for Monte Carlo Simulation (2)

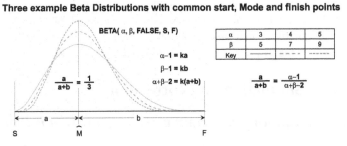

Figure 3.28 Three Example Beta Distributions with Common start, Mode and Finish Points

Table 3.10 Choosing Beta Distribution Parameters for Monte Carlo Simulation Based on the Mode

α		1.5	2	2.5	3	3.5	4	4.5	5	5.5	6
α - 1		0.5	1	1.5	2	2.5	3	3.5	4	4.5	5
β	β - 1										
1.5	0.5	50%	67%	75%	80%	83%	86%	88%	89%	90%	91%
2	1	33%	50%	60%	67%	71%	75%	78%	80%	82%	83%
2.5	1.5	25%	40%	50%	57%	63%	67%	70%	73%	75%	77%
3	2	20%	33%	43%	50%	56%	60%	64%	67%	69%	71%
3.5	2.5	17%	29%	38%	44%	50%	55%	58%	62%	64%	67%
4	3	14%	25%	33%	40%	45%	50%	54%	57%	60%	63%
4.5	3.5	13%	22%	30%	36%	42%	46%	50%	53%	56%	59%
5	4	11%	20%	27%	33%	38%	43%	47%	50%	53%	56%
5.5	4.5	10%	18%	25%	31%	36%	40%	44%	47%	50%	53%
6	5	9%	17%	23%	29%	33%	38%	41%	44%	47%	50%

For the Formula-philes: Selecting Beta Distribution parameters based on the Mode

Consider a Beta Distribution with parameters α and β, a start point S and finish point F and a mode, $\hat{M}$.

The Mode $\hat{M}$ of the Beta Distribution occurs at:

$$\hat{M} = S + \frac{(\alpha - 1)}{(\alpha + \beta - 2)}(F - S) \qquad (1)$$

Re-arranging (1):

$$\left(\frac{\hat{M} - S}{F - S} \right) = \frac{(\alpha - 1)}{(\alpha - 1) + (\beta - 1)} \qquad (2)$$

... which expresses the Mode as a proportion of the Range of a Beta Distribution as the ratio of the alpha parameter less one in relation to the sum of the two parameters less one each.

In conclusion, if we don't know what an actual input distribution is, but we have a reasonable idea of the Optimistic, Most Likely and Pessimistic values, then Monte Carlo Simulation will be quite forgiving if we opt for a rather simplistic triangular distribution instead of a Normal or Beta Distribution. It will not, however, be quite so accommodating to our fallibilities if we were to substitute a uniform distribution by a triangular distribution, or use a uniform distribution where the true distribution has a distinct Most Likely value.

3.2 Monte Carlo Simulation and correlation

3.2.1 Independent random uncertain events – How real is that?

If only using Monte Carlo Simulation for modelling cost or schedule was always that simple, i.e. that all the variables were independent of each other. We should really ask ourselves whether our supposedly independent events are in fact truly and wholly independent of each other.

For instance, let's consider a Design and Manufacture project:

- If the Engineering Department comes up with a design that is difficult to make, then there will be a knock-on cost to Manufacturing
- If Engineering is late in completing their design, or makes late changes at the customer's request, there is likely to be a knock-on effect to Manufacturing's cost or schedule
- If Engineering produces a design on time that is easy to make, with no late customer changes, then that doesn't mean that Manufacturing's costs will be guaranteed Manufacturing is just as capable as Engineering of making their own mistakes and incurring additional costs
- Changes to the design of systems to incorporate customer changes, or in response to test results, may have an impact on other parts of the engineered design, previously frozen
- Programme Management in the meantime is monitoring all the above and their costs may increase if they have to intervene in order to recover performance issues

In essence there is a 'soft link' between various parts of an organisation working together to produce a common output. We can express the degree to which the performance of different parts of the organisation, or the ripple effect of design changes through the product's work breakdown structure are interlinked by partial correlation. We may recall covering this in Volume II Chapter 5 with the aid of the Correlation Chicken, Punch and Judy and the rather more sophisticated Copulas.

To see what difference this makes let's consider our previous example from Section 3.1.6 and impose 100% correlation between the ten cost elements in Table 3.2.

Of course, this is an extreme example and it is an unlikely and unreasonable assumption that all the input cost variables are perfectly positively correlated with one another, i.e. all the minima align and all the maxima align, as do all the corresponding confidence levels in-between. However, it serves us at present to illustrate the power of correlation, giving us the result on the right hand side of Figure 3.29; the left hand side represents the independent events view which we generated previously in Figure 3.19. Table 3.11 summarises and compares the key statistics from the two simulations.

The difference in the results is quite staggering (*apart from looking like hedgehog roadkill! Note: No real hedgehogs were harmed in the making of this analogy, and I hope you weren't eating*), but when we analyse what we have told our Monte Carlo Model to do, it is quite rational. When we define tasks within a model to be perfectly (i.e. 100%) correlated, we are forcing values to match and team up in the order of their values. This is referred to as Rank Order Correlation, which as we know from Volume II Chapter 5 is not the same generally as Linear Correlation.

This gives us three notable features:

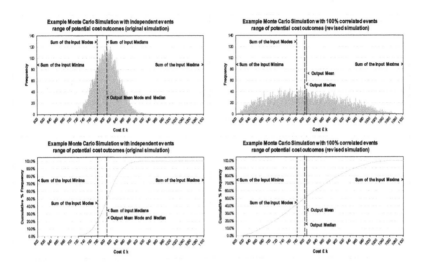

Figure 3.29 Monte Carlo Simulation–Comparison Between Independent and 100% Correlated Tasks

Table 3.11 Comparison of Monte Carlo Simulations with 100% and 0% (Independent) Correlated Events

	Original Simulation		100% Correlation		
	Value £k	Confidence Level	Value £k	Confidence Level	Observation
Minimum	686	0.01%	603	0.01%	Reflects that all optimistic values occur together
Median	809	50.00%	803	50.00%	Median is closer to the Mode in the Ratio 2:1
Mode	809	48.84%	791	45.53%	Mode moves closer to the Sum of the Input Modes
Mean, μ	809.6	50.74%	809.560	52.45%	Sum is stable and tends to the Sum of the Input Means
Maximum	941	100.00%	1076	100.00%	Reflects that all pessimistic values occur together
Left Range	123		188		Reflects change in the Minimum and Mode
Right Range	132		285		Reflects change in the Maximum and Mode
Total Range	255		473		Reflects movement in Minimum and Maximum
Std Dev, σ	35.7		102.9		Standard Devation reflects the wider range
μ - 3σ	702.6	0.06%	500.9	#N/A	Value is less than the absolute Minimum
μ + 3σ	916.6	99.86%	1118.2	#N/A	Value is greater than the absolute Minimum
6σ Range	214.1	99.81%	617.3	#N/A	Value exceeds the absolute Input Range
μ - 2σ	738.2	2.11%	603.8	0.01%	
μ + 2σ	881.0	97.78%	1015.3	97.95%	
4σ Range	142.7	95.67%	411.6	97.94%	Normal Distribution would be 95.45%
Skew	0.050		0.210		Normal Distribution would be zero
Kurtosis	-0.116		-0.765		Normal Distribution would be zero

1. It reduces the frequency of central values and increase occurrences of values in the leading and trailing edges … data is squashed down in the middle and displaced out to the lower and upper extremes, i.e. all the low level values are forced to happen together, and the same for all the high level values
2. The Monte Carlo Output will be less 'Normalesque' in shape and instead will take on the underlying skewness of the input data

3.   The Monte Carlo minimum and maximum values now coincide with the sum of the input minima and the sum of the input maxima respectively

From a pragmatic perspective the likelihood of our getting a situation where all (or some might say 'any') input variables are 100% correlated is rather slim, perhaps even more improbable than a situation in which we have full independence across multiple input variables. In reality, it is more likely that we will get a situation where there is a more of an 'elastic' link between the input variables; in other words, the input variables may be partially correlated. Based on the two extremes of our original model (which assumed completely independent events or tasks), and our last example in which we assumed that all the tasks were 100% Correlated, then intuitively we might expect that if all tasks are 50% Correlation with each other, then the range of potential outcomes will be somewhere between these two extremes

For the Formula-phobes: Impact of correlating inputs

The impact of correlating distribution values together is to move them down and out towards the edges. Let's consider just two input variables. If they are independent, there is a reduced chance of both Optimistic values occurring together, or both Pessimistic values occurring together. As a consequence, we are more likely to get middle range values.

If we set the two variables to be 100% correlated with each other, then all the low-end values will occur together, as will all the high-end values.

There will be no mixing of low and high values which reduces the number of times values occur in the middle range.

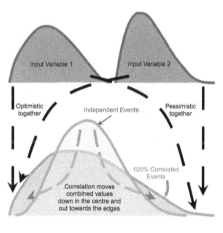

of 0% and 100% correlation. Whether it is 'halfway' or not remains to be seen, but to some extent it depends on how we construct the model and apply the correlation relationship.

3.2.2 Modelling semi-independent uncertain events (bees and hedgehogs)

Does it really matter if we ignore any slight tendency for interaction between variable values? According to the considered opinion and research of the renowned cost estimating expert Stephen Book (1997), the answer would be a definite '*Yes*'. Whilst he acknowledges that establishing the true correlation between variables is very difficult, he advises that any 'reasonable' non-zero correlation would be better (i.e. closer to reality) than ignoring correlation, and by default, assuming it was zero.

As we have already demonstrated, if we have a sufficient (not necessarily large) number of independent cost variables, then the output uncertainty range can be approximated by a Normal Distribution, and that we can measure the variance or standard deviation of that output distribution. Book (1997, 1999) demonstrated that if we were to take a 30-variable model and assume a relatively low level of correlation of 20% across the board, then we are likely to underestimate the standard deviation by about 60% in comparison with a model that assumes total independence between variables. As a consequence, the range of potential outcomes will be understated by a similar amount. Book also showed that the degree of underestimation of the output range increases rapidly with small levels of correlation, but that rate of underestimation growth slows for larger levels of correlation (the curve would arch up and over to the right if we were to draw a graph of correlation on the horizontal access and percentage underestimation on the vertical scale).

Based on this work by Book, Smart (2009) postulated that a background correlation of 20%, or possibly 30%, is a better estimate of the underlying correlation between variables than 0% or 100%, and that the level of background correlation increases as we roll-up the Work Breakdown Structure from Hardware-to-Hardware level (at 20%), Hardware-to-Systems (at 40%) and finally Systems-to-Systems Level integration (at 100%).

Let's explore some of the options we have to model this partial correlation.

However, in order to demonstrate what is happening in the background with partially correlated distributions, a more simplistic logic using a relatively small number of variables may be sufficient to illustrate some key messages on what to do, and what not do, with our Commercial Off-the-Shelf toolset.

Caveat augur

Whilst it is possible to create a Monte Carlo Simulation Engine in Microsoft Excel with partially correlated variables, it is not recommended that we do this without the aid of a dedicated Monte Carlo Add-in, or instead that we use a discrete application that will interact seamlessly with Microsoft Excel. Doing it exclusively in Microsoft Excel creates very large files, and the logic is more likely to be imperfect and over-simplified.

Unless we skipped over Volume II Chapter 5 as a load of mumbo-jumbo mathematics, we might recall that we defined Correlation as a measure of linearity or linear dependence between paired values, whereas Rank Correlation links the order of values together. From a probability distribution perspective, the latter is far more intuitive than the former in terms of how we link and describe the relationship between two variables with wildly different distributions. This is illustrated in Figure 3.30, in which the left hand graph shows two independent Beta Distributions. By forcing a 1:1 horizontal mapping of the Confidence Levels between the two, we can force a 100% Rank Correlation, but as shown in the right hand graph, this is clearly not a Linear Correlation.

Notes:

1. If we took more mapping points between the two distributions then the Linear Correlation would increase as there would be proportionately more points in the central area than in the two extreme tails, giving us a closer approximation to a straight line … but we'll never get a perfect 100% Linear Correlation.
2. The degree of curvature is accentuated in this example as we have chosen oppositely skewed distributions. If the two distributions had had the same positive or negative direction of skew, then the degree of Linear Correlation would increase also.
3. For these reasons, we can feel more relaxed about enforcing degrees of correlation between two distributions based on their Rank Correlation rather than Linear Correlation.

Mapping the distribution Confidence levels together in this way seems like a really good idea until we start to think how should we do it when the distributions are only partially correlated? What we want to get is a loose tying together that allows both variables to vary independently of each other but also apply some restrictions so that if we were to produce a scatter diagram of points selected at random between two distributions

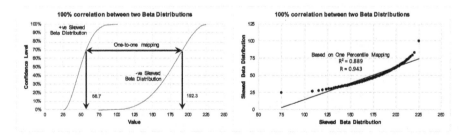

Figure 3.30 Enforcing a 100% Rank Correlation Between Two Beta Distributions

it would look like our Correlation Chicken example from Volume II Chapter 5. We have already seen in that chapter that the Controlling Partner Technique works well to 'push' a level of correlation between an independent and dependent variable, but it doesn't work with semi-independent variables without distorting the underlying input distributions

Iman and Conover (1982) showed that it is possible to impose a partial rank correlation between two variables which maintains the marginal input distributions. Book (1999) and Vose (2008, pp.356–358) inform us that most commercially available Monte Carlo Simulation toolsets use Rank Order Correlation based on the seminal work of Iman and Conover ... but not all; some Monte Carlo toolsets available in the market can provide a range of more sophisticated options using mathematical Copulas to enable partial correlation of two or more variables (Vose, 2008, p.367). Here, in order to keep the explanations and examples relatively simple, we will only be considering the relatively simple Normal Copula as we discussed and illustrated in Volume II Chapter 5.

In Monte Carlo Analysis, when we pick a random number in the range [0,1] we are in essence choosing a Confidence Level for the Distribution in question. The Cumulative Distribution Function of a Standard Continuous Uniform Distribution is a linear function between 0 and 1 on both scales. In Volume II Chapter 5 we discussed ways in which we can enforce a level of linear correlation between two variables that are linearly dependent on each other. However, when we start to consider multiple variables, we must start to consider the interaction between other variables and their consequential knock-on effects. Let's look at two basic models of applying 50% correlation (as an example) to a number of variables:

> A Standard Continuous Uniform Distribution is one that can take any value between a minimum of 0 and a maximum of 1 with equal probability

- We could define that each variable is 50% correlated with the previous variable (in a pre-defined sequence). We will call this '**Chain-Linking**'
- We could identify a 'master' or 'hub' variable with which all other variables are correlated at the 50% level. Let's call this '**Hub-Linking**'

Conceptually, these are not the same as we can see from Figure 3.31, and neither of them are perfect in a logical sense, but they serve to illustrate the issues generated. In both we are defining a 'master variable' to which all others can be related in terms of their degree of correlation. In practice, we may prefer a more democratic approach where all semi-independent variables are considered equal, but we may also argue that some variables may be more important as say cost drivers than others (*which brings to mind thoughts of Orwell's Animal Farm (1945) in which, to paraphrase the pigs, it was declared that there was a distinct hierarchy amongst animals when it came to interpreting what was meant by equality!*).

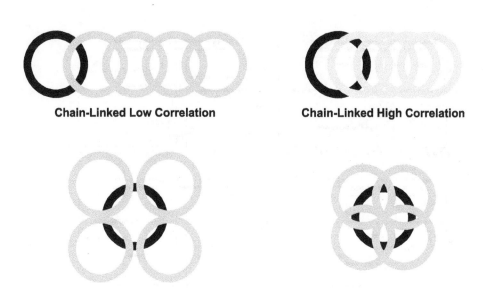

Chain-Linked Low Correlation   **Chain-Linked High Correlation**

Hub-Linked Low Correlation   **Hub-Linked High Correlation**

Figure 3.31 Conceptual Models for Correlating Multiple Variables

3.2.3 Chain-Linked Correlation models

Let's begin by examining the Chain-Linked model applied to the ten element cost data example we have been using in this chapter. If we specify that each Cost Item in our Monte Carlo Model is 50% correlated to the previous Cost Item using a Normal Copula, then we get the result in Figure 3.32, which shows a widening and flattening of the output in the right hand graph in comparison with the independent events output in the left hand graph. Reassuringly, the widening and flattening is less than that for the 100% Correlated example we showed earlier in Figure 3.29.

In Figure 3.33 we show the first link in the chain, in which we have allowed Cost Item 1 to be considered as the free-ranging chicken, and Cost Item 2 to be the correlated chicken (or chick) whose movements are linked by its correlation to the free-ranging chicken. The left-hand graph depicts the Correlated Random Number generated by the Copula (notice the characteristic clustering or 'pinch-points' in the lower left and upper right where the values of similar rank order have congregated ... although not exclusively as we can see by the thinly spread values in the top left and bottom right corners). This pattern then maps directly onto the two variables' distribution functions, giving us the range of potential paired values shown in the right hand graph of Figure 3.33, which looks more like the Correlated Chicken analogy from Volume II Chapter 5.

Figure 3.34 highlights that despite the imposition of a 50% correlation between the two variables, the input distributions have not been compromised (in this case Uniform and PERT-Beta for Cost Items 1 and 2 respectively). As Cost Item 1 is the Lead Variable, the values selected at random are identical to the case of independent events, whereas for

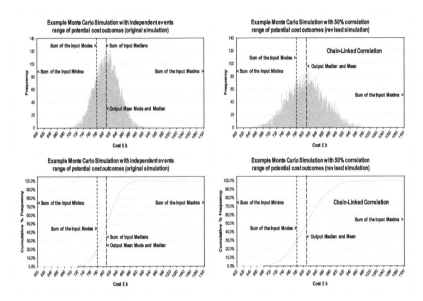

Figure 3.32 Example of 50% Chain-Linked Correlation Using a Normal Copula

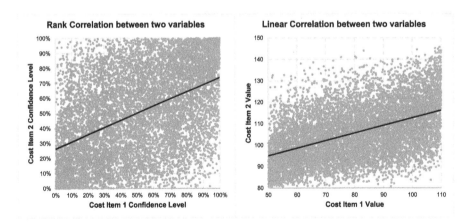

Figure 3.33 Chain-Linked Output Correlation for Cost Items 1 and 2

Cost Item 2, the randomly selected values differ in detail but not in their basic distribution, which remains a PERT-Beta Distribution.

The net effect on the sum of the two variables is to increase the likelihood of two lower values or two higher values occurring together, and therefore decrease the chances of a middle range value to compensate, as shown in Figure 3.35.

This pattern continues through each successive pairing of Chain-Linked variables.

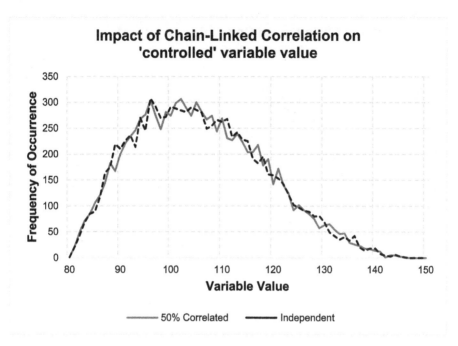

Figure 3.34 Impact of Chain-Linked Correlation on 'Controlled' Variable Value

In Table 3.12 we have extracted the Correlation Matrix which shows that indeed each pair in the chain (just to the right or left of the diagonal, is around the 50% correlation we specified as an input (*which must be re-assuring that we have done it right*). Normally, the diagonal would say 100% correlated for every variable with itself. For improved clarity, we have suppressed these here.) Perhaps what is more interesting is that if we step one more place away from the diagonal to read the Correlation between alternate links in the chain (e.g. first and third or second and fourth) we get a value around 25%. Some of us may be thinking that this is half the value of the chain, but it is actually the product of the adjacent pairs, in this case 50% of 50%.

As we look further away from the diagonal the level of 'consequential' correlation diminishes with no obvious pattern due to the random variations. However, there is a pattern, albeit a little convoluted; it's the Geometric Mean of the products of the preceding correlation pairs (above and to the right if we are below the diagonal, (or below and to the left if we are above the diagonal) as illustrated in Figure 3.36, for a chain of five variables in four decreasing correlation pairs (60%, 50%, 40%, 30%).

Commercially available Monte Carlo Simulation applications are likely to create an error message in these circumstances.

It would also seem to be perfectly reasonable to expect that if we were to reverse the chain sequence then we would get the same Monte Carlo Output. However, in our

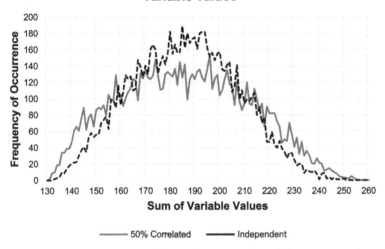

Figure 3.35 Impact of Chain-Linked Correlation on Sum of Variable Values

Table 3.12 Monte Carlo Output Correlation with 50% Chain-Linked Input Correlation

ID	Var 1	Var 2	Var 3	Var 4	Var 5	Var 6	Var 7	Var 8	Var 9	Var 10
Var 1	-	48.6%	25.5%	12.8%	7.0%	5.2%	4.5%	2.9%	1.7%	1.4%
Var 2	48.6%	-	48.9%	25.2%	12.2%	6.5%	2.1%	-0.5%	-0.2%	-0.6%
Var 3	25.5%	48.9%	-	49.9%	24.3%	13.1%	6.5%	1.7%	1.7%	1.4%
Var 4	12.8%	25.2%	49.9%	-	49.5%	24.2%	12.1%	4.0%	2.1%	1.7%
Var 5	7.0%	12.2%	24.3%	49.5%	-	50.1%	24.9%	11.0%	4.3%	3.0%
Var 6	5.2%	6.5%	13.1%	24.2%	50.1%	-	49.7%	22.6%	9.6%	3.9%
Var 7	4.5%	2.1%	6.5%	12.1%	24.9%	49.7%	-	48.6%	23.0%	11.9%
Var 8	2.9%	-0.5%	1.7%	4.0%	11.0%	22.6%	48.6%	-	47.5%	24.9%
Var 9	1.7%	-0.2%	1.7%	2.1%	4.3%	9.6%	23.0%	47.5%	-	50.2%
Var 10	1.4%	-0.6%	1.4%	1.7%	3.0%	3.9%	11.9%	24.9%	50.2%	-

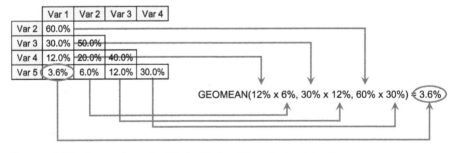

Figure 3.36 Generation of a Chain-Linked Correlation Matrix

Caveat augur

Resist the temptation to close the chain and create a circular correlation loop. This is akin to creating a circular reference in Microsoft Excel, in which one calculation is ultimately dependent on itself.

Commercially available Monte Carlo Simulation applications are likely to create an error message in these circumstances.

example if we were to reverse the linking and use Variable 10 as the lead variable, we would get the results in Figure 3.37 and Table 3.13, which is very consistent with but not identical to that which we had in Figure 3.32. This is due to the fact that we are allowing a different variable to be the lead and the difference is within the variation we can expect between any two Monte Carlo Simulation iterations. The reason for this difference is that we are holding a different set of random values fixed for the purposes of the comparison.

There may be a natural sequence of events that would suggest which should be our lead variable. However, as we will have spotted from this example, chain-linking correlated variables in this way does not give us that background correlation that Smart

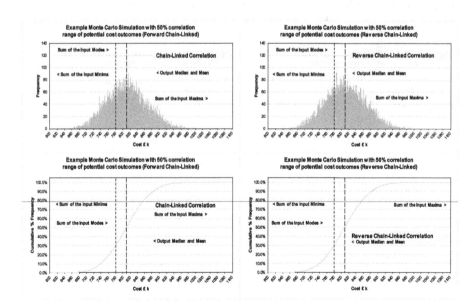

Figure 3.37 Example of 50% Chain-Linked Correlation Where the Chain Sequence has been Reversed

Table 3.13 Output Correlation Matrix where the Chain-Link Sequence has been Reversed

ID	Var 1	Var 2	Var 3	Var 4	Var 5	Var 6	Var 7	Var 8	Var 9	Var 10
Var 1	-	47.8%	25.6%	13.2%	6.9%	4.3%	1.9%	0.1%	0.9%	0.5%
Var 2	47.8%	-	49.4%	25.3%	11.0%	4.2%	0.1%	-1.1%	1.0%	-0.3%
Var 3	25.6%	49.4%	-	50.1%	24.7%	12.3%	6.3%	2.0%	2.5%	1.7%
Var 4	13.2%	25.3%	50.1%	-	49.5%	22.9%	11.0%	4.0%	2.3%	1.6%
Var 5	6.9%	11.0%	24.7%	49.5%	-	48.7%	23.6%	10.3%	4.7%	2.9%
Var 6	4.3%	4.2%	12.3%	22.9%	48.7%	-	49.2%	22.9%	10.8%	5.2%
Var 7	1.9%	0.1%	6.3%	11.0%	23.6%	49.2%	-	49.8%	25.2%	13.2%
Var 8	0.1%	-1.1%	2.0%	4.0%	10.3%	22.9%	49.8%	-	49.2%	25.1%
Var 9	0.9%	1.0%	2.5%	2.3%	4.7%	10.8%	25.2%	49.2%	-	47.9%
Var 10	0.5%	-0.3%	1.7%	1.6%	2.9%	5.2%	13.2%	25.1%	47.9%	-

(2009) recommended as a starting point for cost modelling. Perhaps we will get better luck with a Hub-Linked Model?

3.2.4 Hub-Linked Correlation models

Now let's consider Hub-Linking all the Cost Items to Cost Item 1 (the Hub or Lead Variable). As we can see from the hedgehog plot on the right in comparison with the Mohican haircut on the left of Figure 3.38, Hub-Linked Correlation also widens the range of potential output values in comparison with independent events. A quick look back at Figure 3.32 tells us that this appears to create a wider range of potential outcomes than the Chain-Linked model example.

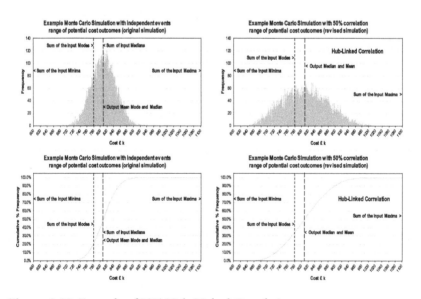

Figure 3.38 Example of 50% Hub-Linked Correlation

Table 3.14 reproduces the Cross-Variable Correlation Matrix (again with the diagonal 100% values suppressed for clarity). Here we can see that each variable is close to being 50% correlated with Hub Variable 1, but the consequence is that all other variable pairings are around 25% correlated (i.e. the square of the Hub's Correlation; we might like to consider this to be the 'background correlation').

Notes:

1) We instructed the model to use a 50% Rank Correlation. Whilst the corresponding Linear Correlation will be very similar in value it will be slightly different. If we want the Output Linear Correlation to be closer to the 50% Level we may have to alter the Rank Correlation by small degree, say 51%.
2) If we were to use an input Rank Correlation of 40%, the background correlation would be in the region of 16%

Let's just dig a little deeper to get an insight into what is happening here by looking at the interaction between individual pairs of values. Cost Items 1 and 2 behave in exactly the same way as they did for Chain-Linking where Cost Item 1 was taken as the Lead Variable in the Chain (see previous Figures 3.33, 3.34 and 3.35). We would get a similar set of pictures if we look at what is happening between each successive value and the Hub Variable. If we compare the consequential interaction of Cost Items 2 and 3, we will get the '*swarm of bees*' in Figure 3.39. We would get a similar position if we compared other combinations of variable values as they are all just one step away from a Hub-Correlated pair ... very similar to being two steps removed from the Chain-Link diagonal.

However, you may be disappointed, but probably not surprised, that if we change the Hub Variable we will get a different result, as illustrated in Figure 3.40 in comparison with previous Figure 3.38. (In this latter case, we used Cost Item 10.) On the brighter

Table 3.14 Monte Carlo Output Correlation with 50% Hub-Linked Input Correlation

ID	Var 1	Var 2	Var 3	Var 4	Var 5	Var 6	Var 7	Var 8	Var 9	Var 10
Var 1	-	48.6%	50.1%	48.9%	48.9%	50.6%	50.4%	49.4%	48.0%	49.5%
Var 2	48.6%	-	24.3%	25.7%	24.3%	25.1%	23.6%	23.2%	24.1%	24.3%
Var 3	50.1%	24.3%	-	26.0%	25.0%	26.9%	26.5%	24.9%	25.4%	26.5%
Var 4	48.9%	25.7%	26.0%	-	25.7%	24.5%	25.3%	23.8%	23.9%	25.5%
Var 5	48.9%	24.3%	25.0%	25.7%	-	26.2%	25.6%	24.4%	23.4%	25.7%
Var 6	50.6%	25.1%	26.9%	24.5%	26.2%	-	26.5%	24.7%	23.9%	24.9%
Var 7	50.4%	23.6%	26.5%	25.3%	25.6%	26.5%	-	25.6%	25.2%	26.7%
Var 8	49.4%	23.2%	24.9%	23.8%	24.4%	24.7%	25.6%	-	24.0%	25.8%
Var 9	48.0%	24.1%	25.4%	23.9%	23.4%	23.9%	25.2%	24.0%	-	25.1%
Var 10	49.5%	24.3%	26.5%	25.5%	25.7%	24.9%	26.7%	25.8%	25.1%	-

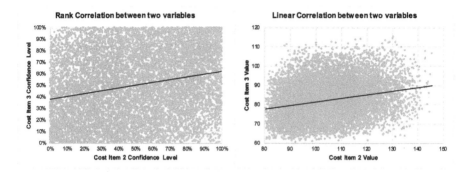

Figure 3.39 Consequential Hub-Linked Correlation for Cost Items 2 and 3

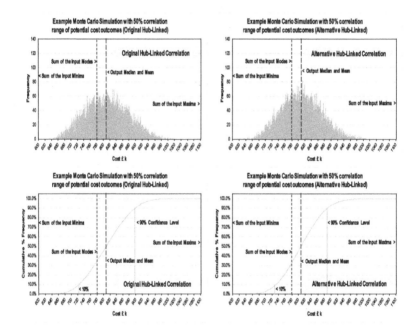

Figure 3.40 Example of 50% Hub-Linked Correlation Using a Different Hub Variable

side, there is still a broader range of potential model outcomes than we would get with independent events. However, it does mean that we would need to be careful in picking our Hub variable. In the spirit of TRACEability (Transparent, Repeatable, Appropriate, Credible and Experientially-based), we should have a sound logical reason, or better still evidence, of why we have chosen that particular variable as the driver for all the others.

Again, using a simple Normal bi-variate Copula as we have here to drive correlation through our model using a specified Lead or Hub-variable, does not give us a constant or Background Isometric correlation across all variables ... but it does get us a lot closer than the Chain-Linked model.

3.2.5 Using a Hub-Linked model to drive a background isometric correlation

Suppose that we find that our chosen Monte Carlo application works as it does for a Hub-Linked model, and we want to get a background Correlation of a particular value, here's a little cheat that can work. Again, for comparison we'll look at achieving a background correlation of around 50%. We'll reverse engineer the fact that the Hub-Linked model above created a consequential correlation equal to the square of the Hub-based Correlation.

1. Let's create a dummy variable with a uniform distribution but a zero value (or next to nothing, so that its inclusion is irrelevant from an accuracy perspective)
2. Use this dummy variable as the Hub variable to which all others are correlated ... but at a Correlation level equal to the square root of the background correlation we want. In this case we will use 70.71% as the square root of 50%

When we run our simulation, we get an output similar to Figure 3.41 with a Correlation Matrix as Table 3.15.

Just to wrap this up, let's have a look in Figure 3.42 at what our background Isometric Correlation of 25% looks like using this little cheat:

• 25% is midway in the starting correlation of 20% to 30% as recommended by Smart (2009), which in turn was based on the on observations made by Book (1997, 1999)
• The Square Root of 25% is 50%, which is what we will use as the correlation level for our Dummy Hub variable

3.2.6 Which way should we go?

The answer is 'Do not build it ourselves in Microsoft Excel.' The files are likely to be large, unwieldly and limited in potential functionality. We are much better finding a Commercial Off-The-Shelf (COTS) Monte Carlo Simulation tool which meets our needs. There are a number in the market that are or operate as Microsoft Excel Add-Ins, working seamlessly in supplementing Excel without the need for worrying whether our calculations are correct. There are other COTS Monte Carlo Simulation tools that are not integrated with Microsoft Excel but which can still be used by cutting and pasting data between them.

How these various toolsets 'push' correlation through the model is usually hidden under that rather vague term '*proprietary information*' but in the views of Book (1999) and

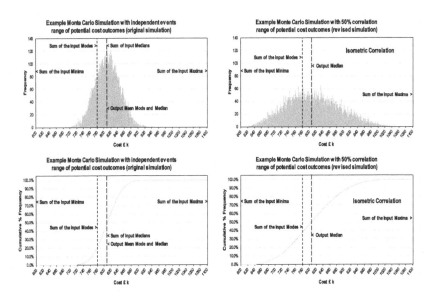

Figure 3.41 Using Hub-Linked Correlation with a Dummy Variable to Derive Isometric Correlation

Table 3.15 Monte Carlo Output Correlation with 50% Isometric Hub-Linked Input Correlation

ID	Var 1	Var 2	Var 3	Var 4	Var 5	Var 6	Var 7	Var 8	Var 9	Var 10	Dummy
Var 1	-	48.5%	49.6%	49.2%	49.3%	50.6%	49.9%	48.7%	48.2%	48.9%	77.4%
Var 2	48.5%	-	50.2%	50.8%	49.9%	51.0%	49.5%	49.2%	49.3%	49.9%	75.0%
Var 3	49.6%	50.2%	-	50.4%	49.7%	51.5%	50.9%	49.6%	49.5%	50.7%	73.4%
Var 4	49.2%	50.8%	50.4%	-	50.6%	50.1%	50.0%	49.1%	48.6%	50.2%	74.8%
Var 5	49.3%	49.9%	49.7%	50.6%	-	51.4%	50.1%	49.5%	48.0%	50.2%	76.5%
Var 6	50.6%	51.0%	51.5%	50.1%	51.4%	-	51.5%	50.2%	49.4%	50.3%	77.6%
Var 7	49.9%	49.5%	50.9%	50.0%	50.1%	51.5%	-	50.2%	49.1%	50.6%	70.8%
Var 8	48.7%	49.2%	49.6%	49.1%	49.5%	50.2%	50.2%	-	49.0%	50.5%	70.2%
Var 9	48.2%	49.3%	49.5%	48.6%	48.0%	49.4%	49.1%	49.0%	-	48.9%	69.1%
Var 10	48.9%	49.9%	50.7%	50.2%	50.2%	50.3%	50.6%	50.5%	48.9%	-	68.8%
Dummy	77.4%	75.0%	73.4%	74.8%	76.5%	77.6%	70.8%	70.2%	69.1%	68.8%	-

Vose (2008) many of them use the algorithms published by Iman and Conover (1982). We should run a few tests to see if we get the outputs we expect:

- Some toolsets allow us to view the Output Correlation Matrix. Does it give us what we expect?
- If a Correlation Matrix is not a standard output, can we generate the correlation between any pairs of variables generated within the model, and plot them on a scatter diagram?

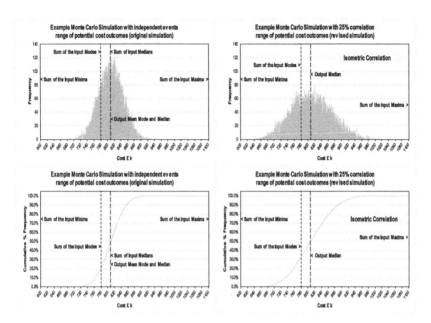

Figure 3.42 Using Hub-Linked Correlation with a Dummy Variable to Derive 25% Isometric Correlation

- Try to replicate the models in this book to see whether we get outputs that are compatible with the Chain-Linked, Hub-Linked or Background Isometric Models.
- Check what happens if we reverse a Chain-Linked model. Does it maintain a consistent output?

Let's compare our three models of Correlation and draw out the differences (Table 3.16). Based on the three Correlation Matrices we have created we would expect that there was a progressive movement of each model away from total independence to total correlation … and not to be disappointed, we have that. (*Don't tell me you were all thinking that I was going to disappoint you?*) The range and standard deviation of the potential outcomes all increase relative to the model to the left. The Mean and the Median remain relatively unfazed by all the correlation going on around them. The 4σ range of a Normal Distribution would yield a Confidence Interval of 95.45%. Our models get less and less 'Normalesque' as we move from left to right, becoming slightly more positively skewed (Skew > 0) and less peaky or more platykurtic (Excess Kurtosis < 0) than a Normal Distribution.

Of one thing we can be sure, whatever model of correlation we choose it will only ever be an approximation to reality, as a result we should choose a model that we can support rationally under the spirit of TRACEability. With this in mind we may want to consider defining the level of Correlation as an Uncertainty Variable itself and model it accordingly. (*We could do this, but are we then not at risk of failing to make an informed judgement and passing the probability buck to a 'black box'?*)

Table 3.16 Comparison of Chain-Linked, Hub-Linked and Background Isometric Correlation Models

	No Correlation		50% Correlation						Full Correlation	
	Independent Events		Chain-Linked		Hub-Linked		Background Isometric		100% Correlation	
	Value	Confidence	Value	Confidence	Value	Confidence	Value	Confidence	Value	Confidence
Minimum	686	0.01%	632	0.01%	635	0.01%	617	0.01%	603	0.01%
Median	809.0	50.00%	809.0	50.00%	809.0	50.00%	806.0	50.00%	803.0	50.00%
Mode	809	48.84%	814	52.91%	834	63.56%	833	61.98%	791	45.53%
Mean m	809.6	50.74%	809.9	50.71%	809.8	50.90%	809.5	51.81%	809.6	52.45%
Maximum	941	100.00%	1000	100.00%	1015	100.00%	1047	100.00%	1076	100.00%
Left Range	123		182		199		216		188	
Right Range	132		186		181		214		285	
Total Range	255		368		380		430		473	
Std Dev s	35.7		55.4		65.9		77.3		102.9	
m - 3s	702.6	0.06%	643.6	0.01%	612.2	0.00%	577.7	0.00%	500.9	0.00%
m + 3s	916.6	99.86%	976.2	99.91%	1007.4	99.98%	1041.3	99.98%	1118.2	100.00%
6s Range	214.1	99.81%	332.6	99.90%	395.2	99.98%	463.6	99.98%	617.3	100.00%
m - 2s	738.2	2.11%	699.0	1.82%	678.1	1.25%	655.0	1.08%	603.8	0.01%
m + 2s	881.0	97.78%	920.8	97.59%	941.6	97.88%	964.1	97.48%	1015.3	97.95%
4s Range	142.7	95.67%	221.7	95.77%	263.5	96.63%	309.1	96.41%	411.6	97.94%
Skew	0.050		0.118		0.116		0.186		0.210	
Ex Kurtosis	-0.117		-0.226		-0.525		-0.493		-0.765	

3.2.7 A word of warning about negative correlation in Monte Carlo Simulation

Until now we have only been considering positive correlation in relation to distributions, i.e. situations where they have a tendency to pull each other in the same direction (low with low, high with high.) We can also in theory use negative correlation in which the distributions push each other in opposite directions (low with high, high with low.)

As a general rule, '**Don't try this at home**', unless you have a compelling reason!

It is true that negative correlation will have the opposite effect to positive correlation where we have only two variables, forcing a Monte Carlo Simulation inwards and upwards, but it causes major logical dilemmas where we have multiple variables as we inevitably will have in the majority of cases.

Caveat augur

Think very carefully before using negative correlation, and if used, use them sparingly.

Chain-Correlation between pairs of variables push consequential correlation onto adjacent pairs of variables based on the product of the parent variables' correlations. This may create unexpected results when using negative correlation.

We can illustrate the issues it causes with an extreme example, if we revisit our Chain-Linked Model but impose a 50% Negative Correlation between consecutive overlapping

Table 3.17 Monte Carlo Output Correlation with 50% Negative Chain-Linked Input Correlation

ID	Var 1	Var 2	Var 3	Var 4	Var 5	Var 6	Var 7	Var 8	Var 9	Var 10
Var 1	-	-50.0%	26.0%	-12.8%	7.3%	-2.3%	3.1%	-0.9%	0.9%	0.2%
Var 2	-50.0%	-	-49.8%	25.7%	-13.6%	5.7%	-5.4%	0.6%	-0.6%	-1.3%
Var 3	26.0%	-49.8%	-	-49.3%	24.7%	-11.2%	7.7%	-3.4%	2.7%	0.3%
Var 4	-12.8%	25.7%	-49.3%	-	-48.4%	22.6%	-12.5%	4.8%	-2.9%	0.6%
Var 5	7.3%	-13.6%	24.7%	-48.4%	-	-48.3%	24.6%	-12.1%	5.3%	-2.1%
Var 6	-2.3%	5.7%	-11.2%	22.6%	-48.3%	-	-48.3%	22.9%	-11.7%	4.8%
Var 7	3.1%	-5.4%	7.7%	-12.5%	24.6%	-48.3%	-	-48.3%	24.0%	-11.0%
Var 8	-0.9%	0.6%	-3.4%	4.8%	-12.1%	22.9%	-48.3%	-	-47.8%	24.7%
Var 9	0.9%	-0.6%	2.7%	-2.9%	5.3%	-11.7%	24.0%	-47.8%	-	-49.1%
Var 10	0.2%	-1.3%	0.3%	0.6%	-2.1%	4.8%	-11.0%	24.7%	-49.1%	-

pairs, (Var 1 with Var 2, Var 2 with Var 3 etc.), giving us an alternating pattern of positive and negative correlated pairs, similar to Table 3.17.

3.3 Modelling and analysis of Risk, Opportunity and Uncertainty

In Section 3.1.3 we defined 'Uncertainty' to be an expression of the lack of sureness around a variable's value that is frequently quantified as a distribution or range of potential values with an optimistic or lower end bound and a pessimistic or upper end bound.

> In principle all tasks and activities will have an uncertainty range around their cost and/or the duration or timescale. Some of the cause of the uncertainty can be due to a lack of complete definition of the task or activity to be performed, but it also includes allowance for a lack of sure knowledge in terms of the level of performance applied to complete the task or activity.
>
> Some tasks or activities may have zero uncertainty, for example a fixed price quotation from a supplier. This doesn't mean that there is no uncertainty, just that the uncertainty is being managed by the supplier and a level of uncertainty has been baked into the quotation.

There will be some tasks or activities that may or may not have to be addressed, over and above the baseline project assumptions; these are Risk and Opportunities. In some circles Risk is sometimes defined as the continuum stretching between a negative 'Threat' and a positive 'Opportunity', but if we were to consult authoritative sources such as the *Oxford English Dictionary* (Stevenson & Waite, 2011), the more usual interpretation has a negative rather than a positive connotation. As we can surmise, this is another of those annoying cases where there is no universally accepted definition in industry. For our purposes in the context of this discussion, we shall define three terms as follows:

Definition 3.2 Risk

A Risk is an event or set of circumstances that may or may not occur, but if it does occur a Risk will have a detrimental effect on our plans, impacting negatively on the cost, quality, schedule, scope compliance and/or reputation of our project or organisation.

Definition 3.3 Opportunity

An Opportunity is an event or set of circumstances that may or may not occur, but if it does occur, an Opportunity will have a beneficial effect on our plans, impacting positively on the cost, quality, schedule, scope compliance and/or reputation of our project or organisation.

Definition 3.4 Probability of Occurrence

A Probability of Occurrence is a quantification of the likelihood that an associated Risk or Opportunity will occur with its consequential effects. T

Within a single organisation it would be hoped that there would be a consistent definition in use, but when dealing with external organisations (or people new to our organisation) it is worth checking that there is this common understanding.

Using the terms as they are defined here, if a Risk or an Opportunity does occur then it will generally have an associated range of Uncertainty around its potential impact. The thing that differentiates the inherent uncertainties around Risks and Opportunities that may arise, and those around the Baseline Tasks, is that the former has an associated Probability of Occurrence that is less than 100%; Baseline Tasks have to be done.

Each Baseline task or activity has a Probability of Occurrence of 100%, i.e. it **will** happen but we cannot be sure what the eventual value will be.

In the same way, **if** the Risk or Opportunity does happen, then we are unlikely to know its exact impact, hence the uncertainty range.

If the Risk or Opportunity **does not** occur, then there is no uncertainty; the associated value is zero.

Performing an analysis of the Risks, Opportunities and Uncertainties on a project, or programme of projects, is an important element of risk management, but it is often also wider than risk management in that risk management processes in many organisations tend to concentrate on Risks and their counterparts (Opportunities), and the Uncertainties around those, but unfortunately, they often ignore the Uncertainties in the Baseline task. Such Uncertainty still needs to be managed; in fact we can go a step further and say that from a probabilistic perspective **Risks and Opportunities should be modelled in conjunction with the Uncertainty around the Baseline task as a single model** as they are all part of the same interactive system of activities and tasks within a project. (Here, we will not be addressing risk management techniques *per se*, but will be looking at the use of Monte Carlo Simulation to review the total picture of Risks, Opportunities and Uncertainties.)

That said, there will be occasions where we want to model Risk and/or Opportunities in isolation from the general Uncertainties in the Baseline assumption set. If we are doing this as part of a wider risk management activity to understand our exposure, then this is quite understandable and acceptable. However, in order that we understand the true impact on cost or schedule or both, then we must look at the whole system of Risks, Opportunities, Baseline tasks and all the associated Uncertainties 'in the round' to get a statistically supportable position.

We will return to this point in Section 3.3.2.

3.3.1 Sorting the wheat from the chaff

Our Definitions of Risk, Opportunity and Uncertainty may seem very 'black or white' but in reality, when we come to apply them we will find that there are areas that can be described only as 'shades of grey'.

Let's consider the simple task of the daily commute to work through rush-hour traffic. Suppose this tortuous journey has the following characteristics:

1) We usually travel by the same mode of road transport (car, bus, bicycle, motor-bike etc.)
2) We usually set off at around the same time in the morning
3) The shortest route is 15 miles and we normally take around 40 minutes to do the journey depending on traffic and weather. (*Why are you groaning? Think yourself lucky; some people's journeys are much worse!*)
4) There are ten sets of traffic lights, but no bus lanes or dual occupancy lanes (*whoopee, if you're in a car! Boo, if you are in a bus, taxi or on a bicycle*)

5) There are limited opportunities for overtaking stationary or slow-moving traffic, and then that depends on whether there is oncoming traffic travelling in the opposite direction

6) If we get delayed, then we start to encounter higher volumes of traffic due to the school run. (*Lots of big 4x4 vehicles looking for somewhere safe to park or stop – I couldn't possibly comment!*)

7) On occasions, we might encounter a road accident or a burst water pipe in the road

Does that sound familiar? Let's analyse some of the statements here, and whether we should treat them as Baseline Uncertainty, Risk or Opportunity:

- The use of the words 'usually' and 'normally' in (2) and (3) implies that there is uncertainty, or a lack of exactness

- Traffic lights in (4) could be interpreted as 'may or may not' be stopped, which we could interpret as a Risk because a red (or amber) light may or may not occur. However, the same could be said of a green light in that it may or may not occur. In this instance as it must be either one of two outcomes (stop or continue), so we are better treating this as uncertainty around the Baseline task of passing through the lights. This is one of those 'shades of grey' to which we referred

- In (5), the word 'Opportunity' is casually thrown in as it can easily be misinterpreted and used in a wider sense than is intended by our Definition 3.3. For instance, we may have an Opportunity to use a motorbike instead of a car, allowing us to overtake more easily. However, in that context, there is no real 'may or may not' chance probability here, it is more likely to be a digital decision: we decide from the outset whether to use a motorbike or a car ... unless we find that the car is difficult to start, in which case, we have our 'may or may not' element that would constitute an opportunity of using a motorbike. However, here in the context of 'opportunities' in (5) and driving a car, we are better interpreting this as being an integral part of the Optimistic range of values

- Similarly, in (6), the use of 'if' suggests that there is an element of 'may or may not' occur. It seems to be hinting at potential positive correlation between events, but in most practical circumstances we can allow this just to be covered by an extended range of pessimistic values

- The clue in (7) that these are risks are the words 'on occasions we might'. There is a definitive 'may or may not' aspect to these, which if they occur, will create delays in excess of our normal range of travel time Uncertainty

Most of us will have spotted that there is no clear Opportunity in our list as we have defined it. This is because they are often harder to identify than Risks. A possible Opportunity here could be:

We know that there is an alternative shorter route we can take to work which passes the site of a major employer in the area. The route is normally very busy, so we avoid it. We may hear on the local radio that traffic is lighter than usual on this route as the threatened industrial action at the plant is going ahead as last-minute talks had failed to reach a compromise. There is a definitive 'may or may not' here (or in this case 'hear') that creates the opportunity for reducing our travel time. We may or may not hear the local Traffic News on the radio before we reach the point where the two routes divide.

Let's look at an example of the 'shades of grey' we were talking about:

We have already highlighted that the traffic lights (4) fall into this category. In (3) we mention that the journey time can vary depending on the weather. There will always be weather of some description, but extreme weather conditions (in the UK at least) are quite rare. For instance, blizzards and hurricane force winds affecting major towns and cities are quite rare from a probabilistic perspective and are likely only to affect travel at certain times of the year. Even if it were to affect us only once or twice a year, this would be less than a 1% probability. We could model this as a positively-skewed peaky Beta Distribution. Alternatively, we could model it as a Triangular Distribution with a truncated but realistic pessimistic journey time based on typical poor weather for the area and consider the extreme weather conditions as something that may or may not occur – in other words, as a Risk.

Consistency is the key in these situations, and we need to ensure when we 'sentence' our assumptions and variables appropriately and that the decisions we make are recorded accordingly in the Basis of Estimate (BoE). It might help if we think of this as the **ECUADOR** technique

Exclusions, **C**onstants, **U**ncertainties, **A**ssumptions, **D**ependencies, **O**pportunities and **R**isks

An **A**ssumption is something that we take to be true (at least for the purpose of the estimate) which has either an associated value that is taken as a **C**onstant or has a value that is **U**ncertain.

Something that may or may not happen is a **R**isk or an **O**pportunity. Some Risks can be avoided or mitigated by **E**xcluding them or placing a **D**ependency on an external third party. Similarly, some Opportunities can be promoted by placing a **D**ependency on an external third party.

Assumptions lead us to specify Constants and Uncertainties.

Risks and Opportunities may be mitigated by Dependencies or Exclusions, which in turn revises our list of Assumptions.

3.3.2 Modelling Risk Opportunity and Uncertainty in a single model

In order to model Risks and Opportunities in our Monte Carlo Simulation we need to add an input variable for each Probability of Occurrence. (*For fairly obvious reasons we won't be abbreviating this to PoO!*) To this variable we will invariably assign a Bernoulli Distribution – or a Binary 'On/Off' Switch (see Volume II Chapter 4). With this distribution we can select values at random so that for a specified percentage of the time, we can expect the distribution to return 100% or 'On' value, and the remainder of the time it will return 0% or 'Off' value.

Rather than think that we need to have an additional variable for Risks and Opportunities it is more practical to use the same Probability of Occurrence variable for modelling Baseline Uncertainty too. In this case, the Probability of Occurrence will always take the value 100%, i.e. be in a state of 'Always On'.

Let's illustrate this in Table 3.18 by looking back to our simple cost model example from Table 3.2 in Section 3.1.6 to which we have added a single high impact risk (*just so that we can see what happens to our Monte Carlo Plot when we add something that may or may not happen*). For now we will assume the background Isometric Correlation of 25% as we had in Figure 3.42 and add a single Risk with 50% Probability of Occurrence and a Most Likely impact of £ 350 k, an optimistic impact of £ 330 k and a pessimistic value of £ 400 k (in other words a BIG Risk). We will assume a Triangular Distribution for the Risk.

Such is the nature of Monte Carlo Simulation and the slight variances we get with different iterations or runs, the less meaningful it is to talk about precise output values. From now on we will show the Monte Carlo Out as more rounded values (in this case to the nearest £ 10 k rather that £ 1 k. So … **no more hedgehogs!** (*Sorry if some of us found that statement upsetting; I meant it from a plot analogy point of view, not from an extinction perspective.*)

Running our model with this Risk produces the somewhat startling result in Figure 3.43, which shows that our fairly Normalesque Output plot from Figure 3.42 has suddenly turned into a Bactrian Camel. (*Yes, you got it, I'm a man of limited analogy.*)

Table 3.18 Monte Carlo Input Data with a Single Risk at 50% Probability of Occurrence

ID	Cost Element	Optimistic (Minimum) £k	Mode £k	Pessimistic (Maximum) £k	Distribution	Probability of Occurrence
1	Cost Element 1	50	80	110	Uniform	100%
2	Cost Element 2	80	101	150	Beta	100%
3	Cost Element 3	60	80	120	Beta	100%
4	Cost Element 4	20	50	80	Triangular	100%
5	Cost Element 5	30	50	100	Triangular	100%
6	Cost Element 6	75	100	150	Beta	100%
7	Cost Element 7	90	100	120	Beta	100%
8	Cost Element 8	70	80	110	Beta	100%
9	Cost Element 9	35	45	55	Uniform	100%
10	Cost Element 10	90	95	105	Triangular	100%
11	Risk 1	330	350	400	Triangular	50%

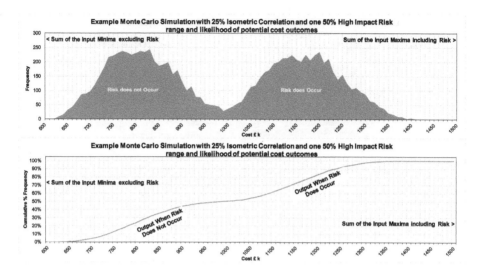

Figure 3.43 25% Isometric Correlation Across Baseline Activities Plus One 50% High Impact Risk

However, if we think of it rationally, it is quite easy to understand what is going on here – it's all a question of Shape Shifting!

- Half the time the Risk does not occur, which gives us the same output shape and values as we had in Figure 3.42 ... but with only half the frequency (left hand hump).
- The other half of the time we get that same shape repeated but with an additional cost value of between £ 330 k and £ 400 k when the Risk occurs ... the shape shifts to the right.
- The knock-on effect to the cumulative Confidence Levels gives us a double S-Curve. As this does not actually go flat at the 50% Level it confirms that two humps do in fact overlap slightly in the middle.

We still have the contraction or narrowing of the potential range but for emphasis we have shown the sum of the Input Minima without the Risk but included the Risk in the sum of the Input Maxima.

If our Risk had only had a probability of occurrence of some 25%, then we would still have the same degree of Shape Shifting, but the first hump would be three times as large as the second. In other words, 75% of the time in our random simulations the Risk does not occur and 25% of the time it does. Figure 3.44 illustrates this; the lower

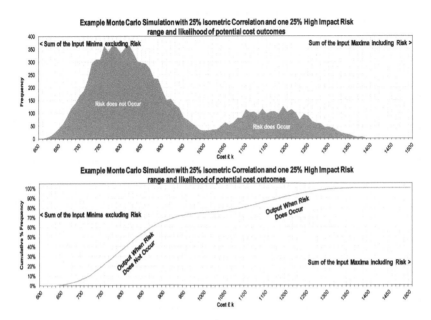

Figure 3.44 25% Isometric Correlation Across Baseline Activities Plus One 25% High Impact Risk

Caveat augur

Note that we have elected not to correlate the Uncertainty around the Risk variable here with any other variable.

If we were to decide to do this (and we have that option) then we should ensure that we are correlating the input distributions prior to the application of the Probability of Occurrence rather than afterwards.

If we have a 10% Risk or Opportunity, then 90% of the time the final output value will be zero.

graph confirms that the first hump accounts for 75% of the random occurrences and the remainder are accounted for by the second hump when the Risk occurs.

It would be a very rare situation that we only had a single Risk to worry about. In fact, it would be bordering on the naïve. Let's see what happens when we add a second Risk at 50% Probability. This time we will define a PERT-Beta Distribution with a range from £ 240 k to £ 270 k around a Most Likely of £ 250 k. Again, we have deliberately exaggerated these values in relation to the Baseline Task to highlight what is happening in the background, as we show in Figure 3.45. (*It resembles something of a rugged mountain terrain.*)

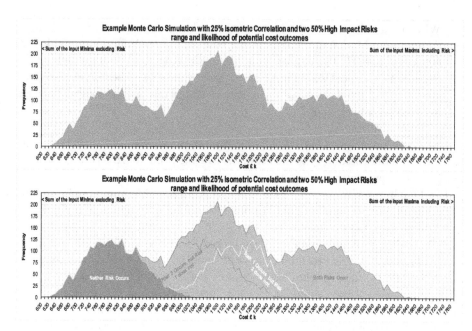

Figure 3.45 25% Isometric Correlation Across Baseline Activities Plus Two 50% High Impact Risks

The upper graph shows the standard Monte Carlo Output; the lower graph attempts to explain what is happening within it. It divides into four quarters from an output frequency perspective:

i.   25% of the time neither risk occurs
ii.   25% of the time Risk 1 occurs but Risk 2 does not
iii. 25% of the time Risk 2 occurs but Risk 1 does not
iv.   25% of the time both risks occur

Each Risk has its own spread of Uncertainty which occurs at random to the Uncertainty ranges around the Baseline tasks but we can verify the approximate position of the second Mode quite simply. (For correctness, we really should be considering the effects of the distribution Means and not the Modes, but Modes are often more accessible for quick validation.):

i.   Baseline Task Central Value occurs around £ 800 k
ii.   We would expect the Nominal Mode for Risk 1 to occur £ 350 k to the right of the Baseline Task Central Value, i.e. £ 800 k + £ 350 k = £ 1,150 k
iii. Similarly, we can expect the Nominal Mode for Risk 2 to occur in the region of £ 250 k to the right of the Baseline Task Central Value, i.e. £ 800 k + £ 250 k = £ 1,050 k

iv. When both risks occur, we can expect the peak to occur some £ 600 k (sum of the two Risk Modes) to the right of the Baseline Central Value, i.e. £ 800 k + £ 600 k = £ 1,400 k. The ultimate peak occurring some £ 300 k to the right of the Baseline Task Central Value can be rationalised as the offset based on the weighted average of the two Risks occurring independently. In this case the weighted average is the simple average of £ 350 k and £ 250 k)

If we had Opportunities instead of Risks, then the principle would be the same only mirrored with the offset being to the left of the baseline condition. Don't forget that we need to enter opportunities as negative costs!

Just think how complicated it gets with several Risks and Opportunities all switching in and out in differing random combinations. (*The good news is that Monte Carlo can cope with it even if we just shake our heads and shiver.*)

Let's look at an example, using our same simple cost model to which we have added four Risks and one Opportunity, as shown in Table 3.19. This gives us the output in Figure 3.46. (We're sticking with the 25% Isometric Correlation.)

As we can see all the risk effects appear to have blurred together in this case (no obvious 'camelling' this time), and gives an apparent shift to the right in relation to the output for the Baseline activities alone, shown as a dotted line. (*Another case of shape shifting, although there is also a good degree of squashing and stretching going on as well.*)

We said earlier that Risks, Opportunities and Baseline task Uncertainties are all part of a single holistic system and should be modelled together … but sometimes modelling

Table 3.19 Monte Carlo Input Data with Four Risks and One Opportunity

ID	Cost Element	Optimistic (Minimum) £k	Mode £k	Pessimistic (Maximum) £k	Distribution	Probability of Occurrence
1	Cost Element 1	50	80	110	Uniform	100%
2	Cost Element 2	80	101	150	Beta	100%
3	Cost Element 3	60	80	120	Beta	100%
4	Cost Element 4	20	50	80	Triangular	100%
5	Cost Element 5	30	50	100	Triangular	100%
6	Cost Element 6	75	100	150	Beta	100%
7	Cost Element 7	90	100	120	Beta	100%
8	Cost Element 8	70	80	110	Beta	100%
9	Cost Element 9	35	45	55	Uniform	100%
10	Cost Element 10	90	95	105	Triangular	100%
11	Risk 1	50	60	70	Uniform	67%
12	Risk 2	65	75	95	Beta	10%
13	Risk 3	40	50	60	Normal	33%
14	Risk 4	55	60	80	Triangular	25%
15	Opportunity 1	-60	-45	-40	Triangular	20%

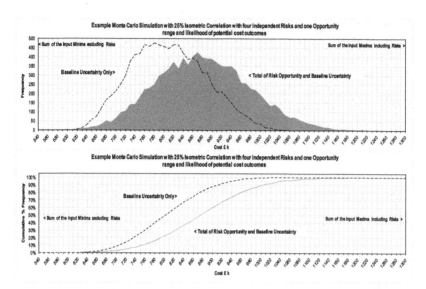

Figure 3.46 25% Isometric Correlation Across Baseline Activities with Four Risks and One Opportunity

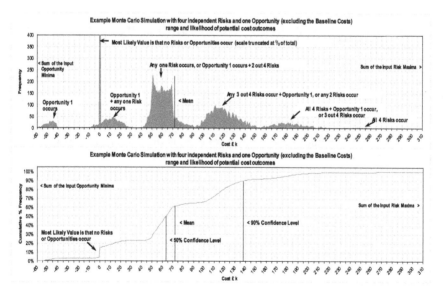

Figure 3.47 Monte Carlo Output for Four Independent Risks and One Opportunity (not Baseline Tasks)

Risks and Opportunities separately from Baseline uncertainties can help us to understand what the 'Background Risk' is, and which risks we simply cannot ignore as background noise – the so-called low probability, high impact risks that we will discuss in

Section 3.3.5 on 'Dealing with high probability Risks'. We may indeed choose to ignore them as a conscious deliberate act, but it shouldn't be left as a by-product of, or as 'collateral damage' from a full statistical analysis.

Figure 3.47 illustrates that by modelling all the Risks and Opportunities in isolation from the Baseline tasks we can begin to understand the key interactions by looking at the combinations of Most Likely Values of the Risks and Opportunities. We have highlighted some of the possible combinations as an example. Assuming that we are working in a risk-sharing environment between customers and contractors then, as it stands with this model, we would probably want to cover the main 'humps' in the profile. From a risk protection perspective, if we are the producer or contractor, we may want to choose a Risk Contingency of some £ 140 k depicting the 90% Confidence Level. However, this may make us uncompetitive, and from a customer perspective it may be unaffordable or unpalatable; the customer may wish to see a 50:50 Risk Share of some £ 64 k. Incidentally, the overall average simulated value of the four Risks and one Opportunity is £ 71.8 k (as indicated on the graph), whereas the Mode or Most Likely value is zero.

Note that the probability of all four Risks occurring simultaneously is very small ... the last hump is barely perceptible. (*It's in the 'thickness of a line' territory.*)

We'll get back to choosing an appropriate confidence level in Section 3.5 after we have discussed a few more issues that we may need to consider.

3.3.3 Mitigating Risks, realising Opportunities and contingency planning

It is not uncommon in some areas to refer to the product of the Most Likely Value of a Risk and its Probability of Occurrence as the Risk Exposure:

Risk Exposure = Most Likely Risk Impact × Probability of Occurrence

We should point out that this can be assumed to include the equivalent negative values for Opportunities. This is not always the case.

It may be more useful, however, if we were to compare each individual Risk Exposure with the total of the absolute values (ignoring the signs) of all such Risk and Opportunity Exposures, giving us a Risk & Opportunity Ranking Factor for each Risk:

$$\text{Risk \& Opportunity Ranking Factor} = \frac{\text{ABS}(\text{Risk Exposure})}{\text{Sum of all ABS}(\text{Risk Exposures})}$$

Our argument to support this differentiation is that if the Risk occurs, we are exposed to its full range of values. If it doesn't occur, there will be no impact, no exposure. It is really just a question of semantics, but regrettably this factoring of the Most Likely Risk Impact by the Risk's Probability of Occurrence is sometimes used as the basis for calculating a contribution to a Risk Contingency, as we will discuss in Chapters 4 and 5. (*As you can probably tell, I'm not a fan of this technique.*) However, the Risk & Opportunity Ranking Factor can be a useful measure in helping us to prioritise our Risk Mitigation and Opportunity Promotion efforts. (Note that the word 'can' implies that this is not always the case!)

The need to be competitive and to be affordable prompts us to look at ways we can ensure that some of our Risks are mitigated or that our Opportunities are promoted or realised. This usually requires us to expend time, effort and resources (and therefore cost) in order to increase or reduce their impact or increase or reduce the likelihood of their arising, depending on whether we are thinking of opportunities or risks. We have to invest to make a potential saving.

To help us make the appropriate decision on where we might be better concentrating our thoughts and efforts, we might want to consider the following using Table 3.20.

- By digitally de-selecting one Risk or Opportunity at a time, we can perhaps see where we should prioritise our efforts.
- We can begin with the Risk or Opportunity which has the highest absolute Risk and Opportunity Ranking Factor (i.e. ignoring negative signs created by Opportunities) and assess the overall impact on the Risk Profile if we can mitigate that 'largest' risk or promote the largest opportunity.
- Figure 3.48 gives an example of Mitigating Risk 1 which had an assumed Probability of Occurrence of some 67%. What it doesn't show is the cost of mitigation, i.e. the investment which will have to be included in a revised list of Baseline tasks when we run our new overall Monte Carlo Simulation of Risk Opportunity and Uncertainty combined.

Here, if we were to mitigate Risk 1, the resulting Risk Profile would suggest that there was now a 47% chance that we would not be exposed to further risk impact. That sounds like a good decision, but potentially, this could give us a real problem if we are in an environment in which it is custom and practice to have a 50:50 Risk Share with a customer or client, based blindly on Monte Carlo Simulation Confidence Levels, then:

Table 3.20 Risk Exposure and Risk and Opportunity Ranking Factor

ID	Cost Element	Optimistic (Minimum) £k	Mode £k	Pessimistic (Maximum) £k	Distribution	Probability of Occurrence	Risk Exposure £k	Risk & Opportunity Ranking Factor
11	Risk 1	50	60	70	Uniform	67%	40.2	46%
12	Risk 2	65	75	95	Beta	10%	7.5	9%
13	Risk 3	40	50	60	Normal	33%	16.5	19%
14	Risk 4	55	60	80	Triangular	25%	15	17%
15	Opportunity 1	-60	-45	-40	Triangular	20%	-9	10%

	Total Absolute Values	88.2	100%

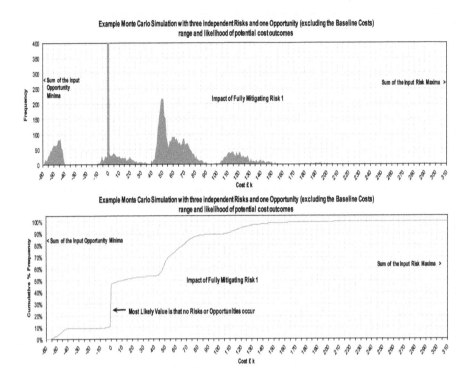

Figure 3.48 Monte Carlo Output Post Mitigation of Risk 1

- A revised 50% Confidence Level of around £ 10 k, which would be woefully inadequate to deal with any of the remaining risks unless we can use it to mitigate them fully or promote the realisation of the opportunity.
- It could lead to inappropriate behaviours such as not mitigating the Risk to ensure a better 50:50 contingency arrangement with a customer!

However, suppose in that situation, prior to any mitigation activity we had a discussion (*that's a euphemism for negotiation*) with the customer/client, and that we were offered a Risk Contingency of some £ 64 k, being approximately the 50% Confidence Level for the Risks and Opportunities shown in Figure 3.48. Furthermore, let's assume that this Contingency is there also to fund any Risk Mitigation activities on the basis that any Risk Mitigation and Opportunity Promotion should be self-funding in terms of what they are trying to avoid or realise.

We can show (Figure 3.49) from our Monte Carlo Model that there is an almost 60% chance that a maximum of only one Risk, i.e. including the probability that no Risks occur. (We have not included the opportunity here.)

Armed with this knowledge, should any one of our Risks occur (and assuming that none of the budget has been spent on mitigation or promotion) then what can we say about a level of contingency set at £ 64 k?

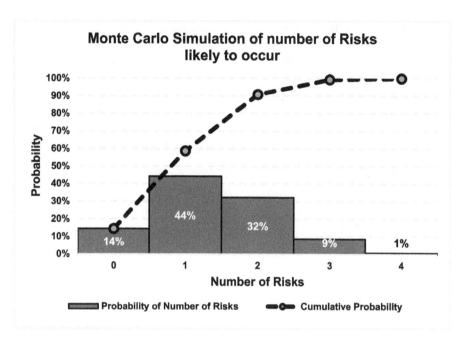

Figure 3.49 Monte Carlo Simulation of Number of Risks Likely to Occur

- It adequately covers the pessimistic range of Risk 1, which has the greatest probability of occurring, with a little left over for other risk mitigation or opportunity promotion. We might conclude that this supports a 50:50 Risk Sharing culture
- £ 64 k is woefully inadequate for Risk 2 … even for the optimistic outcome. Perhaps this is where we would be better investing our time and resource into mitigation activities. (It's going to hurt if it happens, exceeding our 50:50 Risk Share budget and suggesting that a 50:50 Risk Share based purely on Monte Carlo Confidence Levels alone can be somewhat delusional)
- It should cover Risk 3 comfortably as it did for Risk 1 but not if Risk 1 also occurs. Risk 3 is the second most probable to occur and is ranked second using the Risk & Opportunity Ranking Factor
- In terms of Risk 4, the contingency just about covers the range of values less than the Median, given that the Median is between the Mode at £ 60 k and the Mean at £ 65 k
- The Opportunity can boost our Risk Contingency. Let's hope it occurs to allow us to increase our Contingency to cover Risk 2

Using this analytical thought process and the profile in Figure 3.47, we may conclude that a Contingency of around £ 87 k would be more appropriate in order to cover Risk 2 should that be the 'one' to arise, or perhaps around £ 65 k plus the cost of mitigating Risk 2.

Some organisations choose to allocate a Risk Contingency Budget to individual Risks. In this case we would probably allocate it all to Risk 1 (and keep our fingers crossed that none of the others occur). There is a misconception by some, however, that Monte Carlo Simulation will enable us to identify a Contingency for each risk, implying an Expected Value Technique that we will discuss in Section 3.6, which as we can see is somewhat less meaningful when used blindly in the context of Contingency planning.

3.3.4 Getting our Risks, Opportunities and Uncertainties in a tangle

Suppose we were to treat a Baseline Uncertainty as one Risk and one Opportunity rather than as a continuous single variable. Does it matter?

Let's return to our example of a daily commute to work. Let's suppose that we are not on flexible working times and are expected to begin our shift by a certain time. We might say:

• There is a Risk that we will be late caused by our setting off later than we would consider to be ideal

• … but, there is an Opportunity that we will be early for work as a result of our setting off earlier than we normally need

---

For the Formula-phobes: Splitting Baseline Uncertainty into a Risk and Opportunity pair

Consider a Triangular Distribution. Let's split the triangle into two right-angled triangles by dropping a line from the Apex to the Base. The left hand triangle has an area of L% and the right hand triangle has an area of R%. The sum of the two areas must equal the original 100%.

We can consider the left hand triangle as an Opportunity to outturn at less than the Most Likely value with a probability of occurrence of L%. Conversely, the right hand triangle can be thought of as a Risk of being greater than the Most Likely value with a probability of occurrence of R%. However, the two cannot both occur at the same time so any model would have to reflect that condition, i.e. the probability of being less than the Most Likely is the probability of not being greater than the Most Likely (R% = 1 − L%).

For the Formula-philes: Splitting Baseline Uncertainty into a Risk and Opportunity pair

Consider a Triangular Distribution with a range of a to b and a mode at $\hat{m}$. Consider also two right-angled triangular distributions with a common Apex and mode at $\hat{m}$ such that the range of one is between a and $\hat{m}$, and the range of the other is between $\hat{m}$ and b.

Let H be the Modal height from the base.
Then for the Area of the Triangular Distribution between a and b is:

$$100\% = \frac{1}{2}(b-a)H \quad (1)$$

Area $L\%$ of the Left Hand Right-angled Triangle is:

$$L\% = \frac{1}{2}(\hat{m}-a)H \quad (2)$$

Dividing (2) by (1) we can express the area of the Left Hand Triangle in relation to the original triangle:

$$L\% = \frac{1}{2}\left(\frac{\hat{m}-a}{b-a}\right) \quad (3)$$

Similarly the area $R\%$ of the Right Hand Right-angled Triangle:

$$R\% = \frac{1}{2}\left(\frac{b-\hat{m}}{b-a}\right) \quad (4)$$

... which expresses the area of the two right-angled triangles in relation to the proportion of their respective bases to the whole.

Table 3.21 Swapping an Uncertainty Range with a Paired Opportunity and Risk Around a Fixed Value

Cost Element	Optimistic (Minimum) £k	Mode £k	Pessimistic (Maximum) £k	Distribution	Probability of Occurrence	
Cost Element 1	50	80	110	Uniform	100%	
Cost Element 2	80	101	150	Beta	100%	
Cost Element 3	60	80	120	Beta	100%	
Cost Element 4	20	50	80	Triangular	100%	
Cost Element 5	50	50	50	Constant	100%	< Constant Value Assumed
Cost Element 6	75	100	150	Beta	100%	
Cost Element 7	90	100	120	Beta	100%	
Cost Element 8	70	80	110	Beta	100%	
Cost Element 9	35	45	55	Uniform	100%	
Cost Element 10	90	95	105	Triangular	100%	
Risk 1	50	60	70	Uniform	67%	
Risk 2	65	75	95	Beta	10%	
Risk 3	40	50	60	Normal	33%	
Risk 4	55	60	80	Triangular	25%	
Opportunity 1	-60	-45	-40	Triangular	20%	
Cost Element 5 Opp	-20	0	0	Triangular	29%	< Paired Opportunity and Risk to describe
Cost Element 5 Risk	0	0	50	Triangular	71%	Uncertainty around Cost Element 5

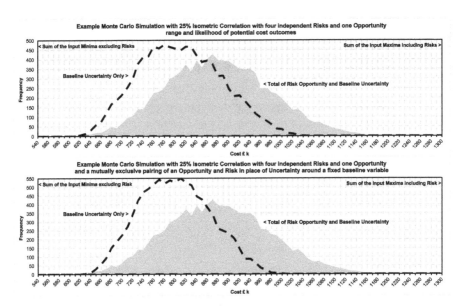

Figure 3.50 Swapping an Uncertainty Range with an Opportunity and a Risk Around a Fixed Value

Equally we could just define our start time as a Baseline variable with an Uncertainty range that spans the range of setting off early through to setting off late, with our Most Likely value being our Normal Setting-off Time.

Let's apply this in our Cost Model example and see what happens. Suppose we replace our Uncertainty range around Cost Element 5 with two right-angled Opportunity 2 and Risk 5 Triangular Distributions as shown in Table 3.21. Cost Element 5 is now assumed to have a fixed value of £ 50 k.

Now let's play a game of 'Spot the Difference' in Figure 3.50.

To be honest it isn't much of a game as there are only two differences:

- Our perception of the range of Baseline Uncertainty is narrower and peakier in the lower graph (i.e. where we have taken Cost Element 5 to be a fixed value, and treated its Uncertainty Range as an Opportunity and a Risk)
- The title of the lower graph has been changed

The more important overall view of Risk Opportunity and Uncertainty is exactly the same in the two graphs, which should give us a nice warm feeling that Monte Carlo Simulation can be somewhat flexible or 'forgiving' if we don't differentiate between Risks and Baseline Uncertainties 'cleanly' ... so long as we don't forget to consider whether there is a corresponding Opportunity for each Risk.

This leads nicely to the next section ...

3.3.5 Dealing with High Probability Risks

The natural consequence of being able to substitute a Baseline Uncertainty by a corresponding Risk and Opportunity pairing allows us to deal with High Probability Risks or Opportunities differently. For instance, with a High Probability, High Impact Risk, it may be better to assume that the implied effect will occur and therefore include it within the Baseline tasks, and to look at the small complementary probability that it doesn't happen, as an Opportunity (i.e. Complementary Probability = 100% − Probability.)

The only thing that will change in the output from our Monte Carlo Modelling will be our perception of the Baseline Activity. For example, consider a Risk such as the following:

- Optimistic, Most Likely, Pessimistic range of £ 80 k, £ 100 k, £ 150 k
- 80% Probability of Occurrence

This can be substituted by a Baseline Activity with the same range of Uncertainty plus an Opportunity:

- Optimistic, Most Likely, Pessimistic range of -£ 150 k, -£ 100 k, -£ 80 k. (Remember that Opportunities must be considered as negative Risks from a value perspective, acting to reduce cost or schedule duration)
- 20% Probability of Occurrence (i.e. the complementary probability of 80%)

Note that as well as the Probability of Occurrence of the Opportunity being the complement of that for the Risk, the sequence and skew of the Opportunity is the reverse of that of the Risk!

Caveat augur

It is very important that we interpret the output from any Monte Carlo Simulation rather than proceed with blind faith in its prophecy. It will perform the mathematics correctly, but when it comes to Risk and Opportunities … things that may or may not happen … then care must be taken to ensure that we are not setting ourselves up for a fall. This is true in a general sense but it is especially so where we have Low Probability High Impact Risks, …. or as we will term them here, 'Extreme Risks'. Just because something is unlikely to occur, it doesn't mean that it won't.

Beware of the False Prophet from Monte Carlo.
Interpret its prophecy wisely.

Table 3.22 Residual Risk Exposure

Probability of Occurrence	High	**Medium Residual Risk Exposure** Can be dealt with on a 'Swings and Roundabouts' approach with a suitable level of contingency	**High Residual Risk Exposure** Consider as part of Baseline Activity with a complementary Opportunity of it not occurring
	Low	**Low Residual Risk Exposure** Can be dealt with on a 'Swings and Roundabouts' approach or even ignored	**Very High Residual Risk Exposure** Digital Decision Required to exclude Extreme Risk from proposal or to include within Baseline Task
		Low	High

Risk Impact Relative to Baseline

Similarly, if we were to identify an Opportunity with a high Probability of Occurrence, we can also assume that it will occur but that there will also be a Risk that we will fail to realise the benefit.

3.3.6 Beware of False Prophets: Dealing with Low Probability High Impact Risks

Let's return to the question of using Risk Exposures as we discussed in Section 3.3.3. Let's extend the thinking to cover the Residual Risk Exposure and define that to be the Risk Exposure that is not covered by a Factored Most Likely Impact Value:

Residual Risk Exposure = Most Likely Impact $\times$ $(100\% -$ Probability of Occurrence$)$

We can use this then to identify Extreme Risks and allows us then to consider our strategy in dealing with them as illustrated in Table 3.22.

The natural extension to the previous situation of a High Probability High Impact Risk is to consider how we might deal with a Low Probability High Impact Risk, i.e. where we would have a Very High Residual Risk Exposure if we were to rely on a Factored Contingency Contribution. The clue to this answer is in Section 3.3.3.

The issue that these types of Risk create can be summarised as follows:

i.   The low probability, high impact values in Monte Carlo Simulations are characterised by a long trailing leg in the Monte Carlo Output with little elevation in relation to the 'bulk' of the output

ii.   If we ignore these Risks by accident (*and this is easy to do if we are not looking for them*) then the consequence of their coming to fruition can be catastrophic

iii. If we ignore these Risks consciously (*on the misguided basis that they will be covered by a general Risk Contingency*) then the consequence of their coming to fruition can be catastrophic as any Contingency is very likely to be inadequate (High Residual Risk Exposure)

iv.   If we assume that the Risk will occur then we will be fundamentally changing our perception of the Baseline task. This may make us unaffordable or uncompetitive. We should also include a complementary Opportunity at 95% Probability that the activity will not be required

v.   If we choose to exclude the Risk from our Estimate (*like the Ostrich burying its proverbial head in the sand*), then it does not make the Risk go away. If it is formally excluded from a contractual perspective, then it just passes the Risk on to the Customer or Client; again, it doesn't go away. This then needs to be followed up with a joined up adult conversation with the customer as this may be the best option for them, the argument being that:

- If we were to include the Risk then we would have to do so on the basis of (iv), and the customer will be paying for it whether or not it occurs
- If we were to exclude it and the customer carries the Risk, then the customer would only pay for it if the Risk were to occur
- The only additional cost to the customer would be if we were to be able to identify an appropriate Mitigation Plan or Strategy in the future, which by implication may need some level of investment

Let's look at the issues using our Cost Model Example. This time we add an extra Risk with a High Impact and Low Probability:

- Optimistic, Most Likely, Pessimistic Range of £ 200 k, £ 220 k, £ 270 k with a Triangular Distribution
- 5% Probability of Occurrence
- The Residual Risk Exposure (based on Most Likely value) is £ 209 k which is some 25%+ of the Baseline Mean of £ 809 k

In Figure 3.51 the upper graph is a copy of Figure 3.47 but with an extended bottom axis to take account of the Extreme Risk. At first glance we might be fooled into thinking that there is no difference, but on closer inspection we can just detect that the peak of the mole hill in the lower graph with the Extreme Risk is slightly less than that in the upper graph which excluded the Extreme Risk. There is a slight extension to the trailing edge, but this is very difficult to see.

The Skew and Excess Kurtosis statistics do not really help us either. We may recall from Volume II Chapter 3 (*unless we passed on that particular statistical delight*) that Skewness is a measure of the degree to which a distribution has a leading or trailing edge

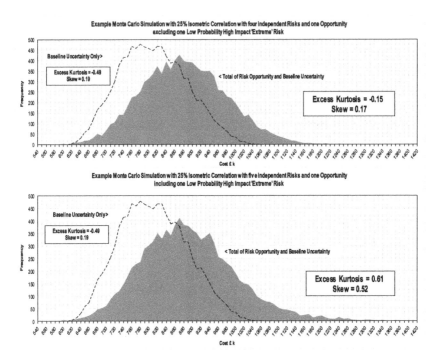

Figure 3.51 Extreme Risks are Difficult to Spot and Easily Overlooked

relative to its Arithmetic Mean, and Excess Kurtosis measures the degree of 'Peakedness' relative to the overall range of values. We might expect that an Extreme Risk increases the positive skewness and the Excess Kurtosis (as the range increases). Well that does happen here, but not to the extent that it would start alarm bells ringing for us as both values including the Extreme Risk are fairly low, i.e. less than one.

> Pearson-Fisher Skewness Coefficient measures the difference in left and right hand ranges relative to the Arithmetic Mean. Some definitions of Skewness are more conveniently expressed in relation to the Median (see Volume II Chapter 3).

However, if we temporarily set aside the dominant Baseline tasks and just look at the Risks and Opportunities, before and after we add the Extreme Risk, then we can see the effects of the Extreme Risk a little more clearly in the lower graph of Figure 3.52 (*if we squint a little*), almost like distant headland further down the coast. (*That's almost poetic; let's hope we don't get all misty eyed; some of us have got that glazed look again already!*) Even now the impact of the Extreme Risk could be easily overlooked as 'noise'.

However, whilst the change in the degree of Skewness is not insignificant, the impact on the Excess Kurtosis is an order of magnitude different. An Excess Kurtosis greater

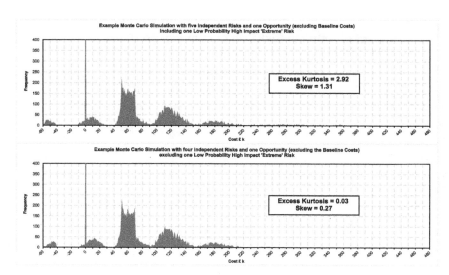

Figure 3.52 Extreme Risks are Easier to Spot in the Context of Risks and Opportunities Only

than 1 should prompt us to look at our Risk Profile more closely. We may then see that distant headland on the coast even if we had missed it on our first visual inspection. The smaller the Probability of Occurrence and the greater the impact, then the greater the Excess Kurtosis will be.

Just to emphasise how easy it is to overlook or dismiss the Extreme Risk, Figure 3.53 compares the Output Confidence Levels of four variations of our cost model. Reading from left to right:

1. Baseline tasks only (no Risks or Opportunities)
2. Baseline tasks plus Risks and Opportunities EXCLUDING the Extreme Risk
3. Baseline tasks plus Risks and Opportunities INCLUDING the Extreme Risk
4. Baseline tasks plus Risks and Opportunities PLUS the Extreme Risk included as a Baseline task (i.e. assuming that we can neither mitigate it nor negotiate its exclusion from the contract, so we assume that it will occur?)

If we are honest with ourselves, at first glance we would probably make a very similar recommendation on an estimate value based on the middle two profiles. We wouldn't feel comfortable reading simply from the left hand line. No-one in their right minds would take the right hand line … at least not until we had explored all the options to mitigate it, which includes that difficult adult conversation with the customer/client.

We could say that the principles outlined here apply equally to Extreme Values of Opportunity. Theoretically this is true, but pragmatically it is much less evident due to the natural positive Skewness of cost and elapsed time as we discussed in Section 3.1.7.

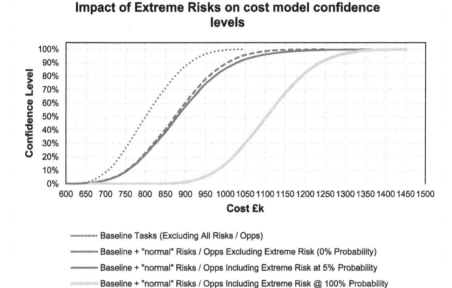

Figure 3.53 Impact of Extreme Risks on Cost Model Confidence Levels

3.3.7 Using Risk or Opportunity to model extreme values of Uncertainty

Based on our discussion in Section 3.3.4, we can choose to segregate extreme values of Baseline Uncertainty from the values that are more pragmatic considerations from an estimating or scheduling perspective. However, from our deliberations in Section 3.3.6, we may conclude '*What's the point … they are not going to show up anyway in any materially significant way!*' The point is that it allows us to remove the extremes from our consideration … or better still identify them so that they can be mitigated in some way, without us having to place restrictions on the more general range of values (some might say that is just being realistic, but 'being pragmatic' is nearer the truth). We may then choose to insure ourselves from some of the extremes and their consequential detrimental effects. It also enables us to avoid some endless (and to some degree pointless) arguments about whether something is a Risk or just a Baseline Uncertainty.

Talking of 'whether or not', let's consider the example of the weather. (*Was that you groaning, or just your stomach rumbling?*) We will always have weather, so that implies that it should be treated as a Baseline Uncertainty, but different extremes of weather could be considered depending on the time of year as Risks or Opportunities. For instance:

- What is the chance of snow in summer in the south of England?
- Very, very small, we would probably agree; yet it snowed in London on 2nd June 1975 (*I was there! I thought I was delusional, that my Maths Exams had finally got to me! Check it out with your favourite Internet Search Engine, if you don't believe me.*)
- How many construction projects at the time had it included as a Risk with potentially disruptive effects? (*It's a rhetorical question, to which I don't know the answer but I assume it to be a round number … 0.*)

By considering, and excluding these extremes of Uncertainty we will not be creating a model with any greater degree of statistical relevance, but we can use these thought processes to place practical limits on what is and is not included. It highlights some things that we need to mitigate that otherwise we don't expect to happen. Augustine's Law of Amplification of Agony (1997, p.301) warns us against that particular folly.

3.3.8 Modelling Probabilities of Occurrence

In the majority of cases, the Probability of Occurrence used is a subjective estimate; one of those created using an Ethereal Approach, often (although not always) using a Trusted Source Method to 'come up with' a Most Likely value of the Probability of Occurrence. In other words, it is largely guesswork, and at best, a considered opinion. It may cross our minds that we should really model it as a variable with an Uncertainty Range somewhere between zero and 100%.

Let's see what happens if we were to assume that the Probability of Occurrence should be modelled as a PERT-Beta Distribution between 0% and 100%, with a Mean Value based on the Single-point Deterministic value which we would choose using the Ethereal Approach and Trusted Source Method as the 'typical' practice. The results are shown in Figures 3.54 and 3.55.

Figure 3.54 looks at the impact on the Overall Risk and Opportunities, and we would probably conclude that it makes very little difference to the outcome. Figure 3.55 confirms our suspicions from a cumulative confidence level perspective. Our conclusion therefore must be '*Why bother with the extra work?*' Needless to say (*but I'll write it anyway*), the impact on the Confidence Levels of the overall Risk, Opportunity and Uncertainty Model will similarly be negligible. From this we might take three messages:

1. Modelling Probabilities of Occurrence with an Uncertainty Range rather than a Single-point Deterministic value, will give us the 'same result' as the modelling with discrete single-point values, provided that …

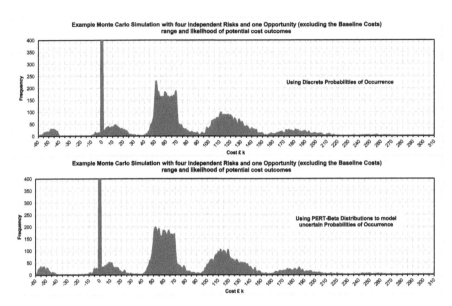

Figure 3.54 Impact of Modelling Probabilities of Occurrence with Uncertainty Distributions (1)

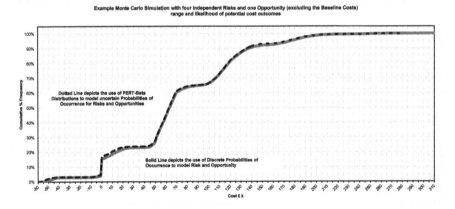

Figure 3.55 Impact of Modelling Probabilities of Occurrence with Uncertainty Distributions (2)

2. The Probability of Occurrence for any Risk or Opportunity reflects the Mean Probability for each **and not** the Most Likely. See Volume II Chapter 4 on the relationships between Mean and Mode for a range of commonly used Distributions.

3. If we truly believe that our discrete Probability of Occurrence is the Most Likely probability, then we should adjust it for modelling purposes to reflect the 5M's as for modelling purposes it will either be under or overstated.

Caveat augur

If we choose to model the Probabilities of Occurrence for Risks and Opportunities, then we must ensure that the Distribution parameters assume that the equivalent Single-point Deterministic values that are typically used represent the Mean values, and not the Most Likely values, which may be our more intuitive starting point.

The net result will be that the effective Probabilities of Occurrence used in the model will be greater than the Most Likely if it is less than 50%, or less than the Most Likely if it is greater than 50%. In other words, modelling returns the effective mean probability, and therefore a single-point deterministic estimate of the Probability of Occurrence should likewise reflect the Mean probability and not the Most Likely.

For the Formula-phobes: Mean in relation to the Mode and Median

Recall the 5M Rule of Thumb from Volume II Chapter 2, the sequence of Mode, Median, Mean changes depending on the direction of skew in our data.

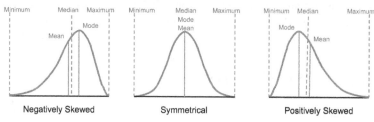

As a consequence, any assumption of the Probability of Occurrence we use should be the Mean value and not the Most Likely. The average of the Random Samples from a Distribution will be approximately equal to the Mean of the Distribution (if the sample size is large enough).

Some organisations' response to this is to limit the range of Probabilities used to a discrete set of values which are aligned with a qualitative assessment/judgement of the Likely Occurrence, as illustrated in the example in Table 3.23. Each organisation will assign the values pertinent to them, and may have a different number of categories.

Table 3.23 Example Probabilities Aligned with Qualitative Assessments of Risk/Opportunity Likelihood

Qualitative Assessment of Likelihood	Probability of Occurrence Used	Implied Range
Very Low	5%	< 10%
Low	20%	10% to 30%
Medium	40%	30% to 50%
High (more likely to occur than not)	65%	50% to 80%
Very High (consider moving to Baseline)	90%	> 80%

3.3.9 Other random techniques for evaluating Risk, Opportunity and Uncertainty

There are other random sampling techniques that can be used to create views of potential outcomes for Risks, Opportunities and Uncertainties such as Latin Hypercube Sampling (LHS) and Bootstrap Sampling techniques.

Don't panic! We're not going through any of these in detail … you've suffered enough already as it is!

These techniques are usually only appropriate where we have access to specialist software to support us. Here we will just give an outline of how they differ from Monte Carlo Simulation (*just in case the name crops up and you think 'Mmm, what's that?'*).

Latin Hypercube Sampling (LHS)

LHS is sometimes packaged with Monte Carlo Simulation software applications and therefore becomes an accessible option. It puts greater emphasis on the 'tails' of the model than we would get through the more commonly used Monte Carlo Simulation. It does this by dividing each variable's range of potential values into intervals of equal probability, e.g. 10 10% or 20 5% intervals. (It assumes the same number of intervals for each variable.)

To help us understand how LHS works, let's first consider the simple model of a Latin Square which has only two variables. A Latin Square is an arrangement of symbols such that each symbol appears once, and only once, in each row and column. For instance, if each variable's values are divided into tertile confidence intervals of nominally 0%–33%, 33%–67%, 67%–100%, we can draw 12 Latin Squares, as illustrated in Figure 3.56. For the two variables, each Latin Square allows us to draw three unique samples, although it can be argued that from a sampling perspective if we were to interchange B and C we would duplicate half the combinations. (Note that the left hand six sets of three samples are equivalent to the right hand six sets if we do this.) In this way we can select 18 unique Latin Square samples for two variables with three tertile confidence intervals.

A Latin Hypercube extends this concept into multiple dimensions. As the number of variables and quantile confidence intervals we choose for each variable increases, so too does the number of unique sample combinations available to us. The Latin Hypercube then forces samples to be taken in the extreme lower and upper ranges of each variable that may otherwise be missed by pure random sampling.

Bootstrap Sampling

Any organisation only has access to a limited amount of data in any real detail, typically its own data based on its experiences and performance. Typically it does not have access to all the detail history of relevant data performed by other organisations in the wider population. There are Software Applications that can be purchased or licensed that come packaged with databases, but these are often limited in the level of detail that can be accessed by the user.

Typically used in conjunction with rather than a replacement for Monte Carlo Simulation, Bootstrap Sampling allows us to choose random samples from an empirical distribution rather than a known or assumed (and theoretically perfect) distribution. As the number of data points available to us is often limited, the Bootstrap Technique requires

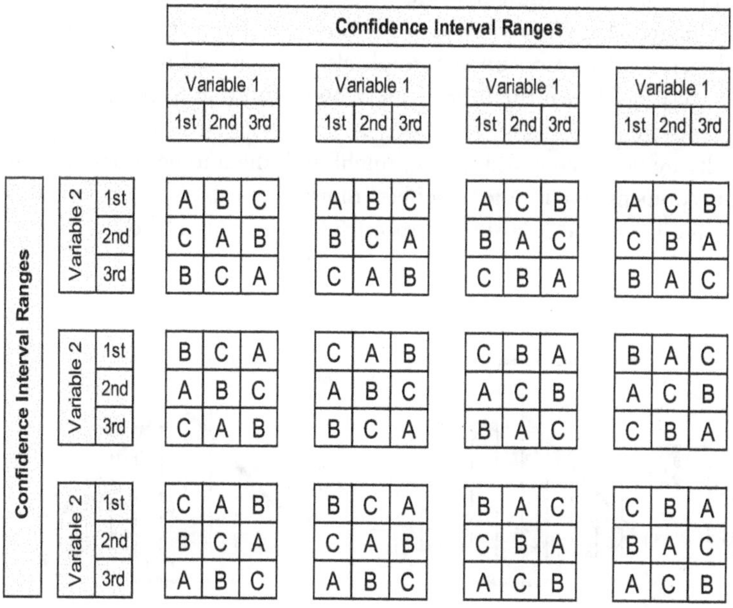

Confidence Interval Ranges											
Variable 1			Variable 1			Variable 1			Variable 1		
1st	2nd	3rd	1st	2nd	3rd	1st	2nd	3rd	1st	2nd	3rd
A	B	C	A	B	C	A	C	B	A	C	B
C	A	B	B	C	A	B	A	C	C	B	A
B	C	A	C	A	B	C	B	A	B	A	C
B	C	A	C	A	B	C	B	A	B	A	C
A	B	C	A	B	C	A	C	B	A	C	B
C	A	B	B	C	A	B	A	C	C	B	A
C	A	B	B	C	A	B	A	C	C	B	A
B	C	A	C	A	B	C	B	A	B	A	C
A	B	C	A	B	C	A	C	B	A	C	B

Confidence Interval Ranges (Variable 2: 1st, 2nd, 3rd for each of three groups)

Figure 3.56 3x3 Latin Square Sampling Combinations for Three Confidence Intervals

us to select sample points at random, record them and then throw them back in the pot to be picked again. Statisticians refer to this as 'Sampling with Replacement'.

To some of us this may sound like a slightly dubious practice, picking data from what may be quite a limited sample. To others amongst us it may make absolute sense to limit our choice to the evidence we have rather than assume or imply a theoretical distribution which may or may not in reality be true, as it is only testing assumptions about what the sample distribution is.

Let's see how it works, and what the implications might be.

Take an example of rolling a conventional six-sided die. Let's suppose that after 25 rolls, we've had enough (*not the most stimulating exercise we've had, was it?*) and our results of our endeavours are shown in the left hand graph of Figure 3.57. There appears to be a bias to the lower three numbers (60%) compared with the upper set of three. We would have expected a more even or uniform distribution but either:

- We haven't rolled the die enough – our sample size is too small
- The die is loaded – it is biased towards the lower numbers

We don't know which is true, but we will use the Bootstrap Sampling technique to get a sample distribution for the sum of adding together the faces of two such dice. If we were to use Monte Carlo Simulation to generate a sample size of 1,080 based on this empirical distribution for one die, we would get the distribution shown in the middle graph of Figure 3.57. For reference, we have superimposed the theoretical Geometric Distribution on the graph also as a dotted line, and in the right hand graph we have shown the more widely used Random Sampling for the Theoretical Distribution.

Whilst the results are not as convincing as we might wish them to be, it does appear to be making some compensation for the imperfect empirical distribution sample, and as such it is better than a 'guess' and at least supports our TRACEability objective.

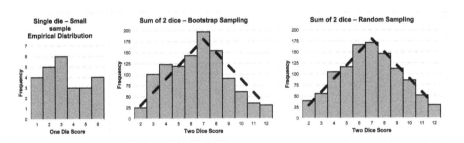

Figure 3.57 Example of Bootstrap Sampling with a Small Empirical Distribution

Expert Judgement

Now the cynics amongst us will probably be saying that the Expert Judgement Technique should fall into this category of 'other random techniques' as it requires guesswork by the Expert. To an extent this is true but it is not entirely random. If the person providing that input is truly an expert, then that person's experience and knowledge will narrow the range of potential values into something that is credible … let's call that 'educated guesswork'.

If we are truly uneasy with using that technique, then we can always ask that expert to provide a lower and upper bound on the Most Likely value provided. Alternatively, we can always ask other Experts for their valued opinions as well. We can always use a pseudo-Delphi Technique to refine their thoughts:

1.   Ask them to provide their views independently at first (using the same brief)
2.   Sit them round the table and ask them to explain their thinking and/or opinion to the other experts to see if they can influence each other to their way of thinking
3.   When they have finished their discussion, use the output as the basis for a range estimate

We can build on this to create an improved profile of how the underlying probability distribution might look, by layering the ranges from each Subject Matter Expert (SME) to create a crude histogram as illustrated in Figure 3.58.

3.4 ROU Analysis: Choosing appropriate values with confidence

There a number of ways in which we can choose a value from the range of potential values we can create from a Risk Opportunity and Uncertainty Analysis:

i.   We could take a structured, reasoned approach to selecting a value based on the scenario in question using all available information
ii. We could take the average of all the values in our Monte Carlo Model
iii. We could ask our granny to pick a number from 1 to 100, and use that to denote the Confidence Level we are going to use from our Monte Carlo Model Output

Well, perhaps not the last one, it's not repeatable unless we know that granny always picks her favourite lucky number, in which case we should have known before we asked. As for the second option, it would appear to follow the spirit of TRACEability but it pre-supposes that Monte Carlo Simulation is always right and dismisses any other approach. (*We'll get back to that point very shortly.*)

Let's take the first option and look at all the information we have to hand, not least of which is that we should not be overly precise with the value we choose as our 'answer' or recommendation. We'll be using the same Cost Model that we used in Section 3.3 as our example.

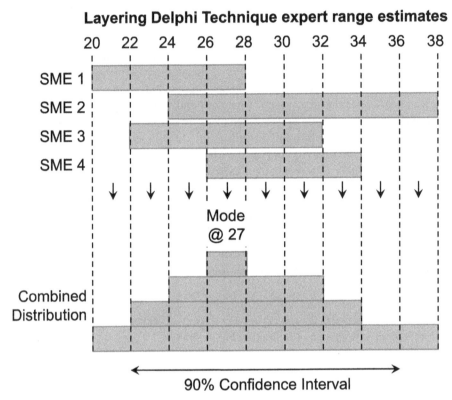

Figure 3.58 Layering Delphi Technique Expert Range Estimates

The intuitive reaction from many of us may be to dismiss the Top-down Approach out of hand as too simplistic, too pessimistic etc. After all, the Bottom-up Approach is more considered, and more detailed, and as the saying goes, '*The devil is in the detail!*' ... well, actually, in this case the devil is NOT in the detail because ...

3.4.1 *Monte Carlo Risk and Opportunity Analysis is fundamentally flawed!*

That got you attention, didn't it? However, it wasn't just a shock tactic, it is in fact true when it comes to a Risk and Opportunity Analysis of cost and schedule.

Monte Carlo Simulation is a beautiful elegant statistical technique (*OK, I'll seek therapy*) but it is fundamentally flawed in that it is almost always incomplete. We can illustrate this quite simply by considering the 2x2 'Known–Unknown Matrix' shown in Table 3.24.

Table 3.24 The Known Unknown Matrix

Maturity of Task Definition and Performance	**Unknown or Poorly Defined**	**Unknown Knowns** We know we have to do the task but its exact scope is not clear, nor do we know how well we will perform the task	**Unknown Unknowns** These are the genuine Risks and Opportunities that exist in reality, but that we have not considered because they haven't occurred to us
	Known or Well Defined	**Known Knowns** We know we have to do the task and we are clear of the requirements and understand our likely performance.	**Known Unknowns** We have identified tasks that may or may not need to be carried out. These are our defined Risks and Opportunities.
		Will Occur	May or May Not Occur

Likelihood of Occurrence

The 'Unknown Unknowns' are those Risks (or potentially Opportunities) that we haven't considered. Our failure to identify them doesn't make them any less real, or less likely to occur, but it does imply that our view of Risks and Opportunities is incomplete. In other words, a 50% Confidence Level in Monte Carlo Simulation is only the 50% Level based on what we have included in our model and not the 50% Level of the true total. As we have already discussed, due to the natural Skewness in our cost or time data, the Unknown Unknown Risks will quite naturally outweigh any Unknown Unknown Opportunities. The same is true for any other Bottom-up Technique. Donald Rumsfeld's 2002 perspective on this is an often-cited quotation.

As a consequence, any Monte Carlo Simulation of Cost or Schedule that purports to consider Risks and Opportunities, is inherently optimistically biased, and therefore is fundamentally flawed, not in terms of what it includes but the fact that it is inherently incomplete; it provides a 'rose-tinted glasses' perspective on life.

Note: We are not talking about Risks or Opportunities that we have consciously excluded and documented as such, just the ones that we haven't considered, or perhaps have dismissed out of hand and not bothered to record.

A word (or two) from the wise?

'There are known knowns. These are things we know that we know. There are known unknowns. That is to say, there are things that we now know we don't know. But there are also unknown unknowns. These are things we do not know we don't know.'

Donald Rumsfeld
United States Secretary of Defense
DoD news briefing
12 February 2002

We can always revisit our list of Risks and Opportunities using a Checklist (as we will discuss for the Top-down Approach in Section 4.1.3) to see whether there are any that we have overlooked … but it will always be prone to being imperfect. Table 3.25 summarises how the Known-Unknown Matrix relates to Monte Carlo Simulation Models in respect of Baseline Tasks, Risks, Opportunities, and the associated ranges of Uncertainties.

Armed with this awareness we have a choice.

1. Dismiss Monte Carlo Simulation as a mere Confidence Trick (*Surely you saw that one coming?*)
2. Make appropriate adjustment or allowance to how we interpret its output for the Unknown Unknowns

The second option is the more useful … but how do we make that adjustment, and to what extent?

The simplest way is to read a higher Confidence Level from our Monte Carlo Output to compensate for the understatement, but by how much? Having done all the detail analysis, we shouldn't just leave this final decision down to guesswork. It would be better to make an informed judgement based on an alternative Approach, Method or Technique. (Enter Stage Left – the 'Slipping and Sliding Technique' as one such way of making an informed, auditable and repeatable decision. We will be discussing this in Chapter 4.)

Despite this fundamental drawback, Monte Carlo Simulation is still worth doing. However, we should note that it is essential that we have a robust Risk and Opportunity Register on

> # A word (or two) from the wise?
>
> *'To know that we know what we know, and that we do not know what we do not know, that is true knowledge.'*
>
> **Confucius**
> Chinese Philosopher
> 551–479 BC

Table 3.25 The Known Unknown Matrix–Monte Carlo View

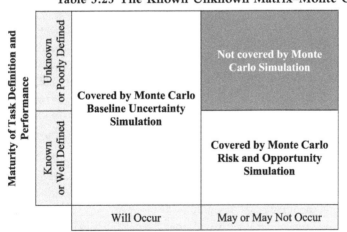

which to base it. Consequently, early in a project, when the Risk and Opportunity Register is somewhat less mature, we should always look to other techniques to validate our conclusions.

3.5 Chapter review

Well, hasn't this chapter been fun? (*It was a rhetorical question!*) We began with a very brief history of Monte Carlo Simulation and where it might be useful to the estimator. It can be used to model a whole variety of different situations where we have a number of variables with a range of potential values that could occur in different combinations. Monte Carlo Simulation is simply a structured technique of using multiple random numbers to generate a not-so-random but stable and consistent profile of the range of potential outcomes. This range is narrower than we would get by simply summating the Optimistic and Pessimistic values, which is down to the probabilistic scenario that not all the good things in life will happen together, nor will all the bad things; it's more of a mixture of the good and the bad together, giving us a result somewhere in the middle. We confirmed that for cost and schedule variables, despite the input variable distributions being skewed (usually positively), the output distribution can usually be approximated by a Normal distribution. However, in other situations the output will not necessarily be 'Normalesque' (*remember the example of the anomalous heights of school children!*). One of the most common uses for Monte Carlo Simulation in estimating is to assess the impact of Risk, Opportunity and Uncertainty.

We demonstrated that we don't have to get the input distributions precisely correct to get a valid Output Distribution; Monte Carlo Simulation is very forgiving ... so long as we get the basic distribution shapes correct, and we don't need to be too precise either; Monte Carlo is a model of accurate imprecision!

However, we also learnt that in many cases when we are considering cost or schedule, we cannot assume that all the input variables are totally independent of each other. Research has shown that a background correlation in the region of 20% to 30% is a more reasonable assumption than total independence. This has the effect of widening the output distribution, pushing down in the middle and outwards to both sides. We then went on to explore different options we can assume in 'pushing' correlation into a model across all its variables such as Chain-Linking and Hub-Linking.

Monte Carlo Simulation is often used to model Risks, Opportunities and Uncertainties. These are consequences that may or may not arise, but if they do, they are either detrimental or beneficial respectively. In order to model Risks and Opportunities we need to add a variable for the Probability of Occurrence for each Risk or Opportunity. The effects of this will tend to create models that are less Normalesque, and could indeed be quite markedly positively skewed due to the effects of Risks and Opportunities 'switching on and off' in the modelling process. Whilst we can model Risk and Opportunities independently of the Baseline tasks to understand the interaction of these, it is essential that the final Monte Carlo Simulation includes all Risks, Opportunities and Baseline task Uncertainties as a single model as they exist together in a single interactive system.

One of the dangers of Monte Carlo Simulation is that we put too much blind faith in the numbers that it generates without thinking about what it is telling us, or is not telling us. It's not that the way that it does the mathematics, is in any way suspect, but those same calculations do tend to mask the impact of Low Probability, High Impact Risks. Unless we select the very high end of the Confidence Range, then this class of Risks will not be covered by a more competitive output value from the model; there will be a high residual Risk exposure in the event (albeit unlikely by definition) of the Risk occurring. We have to make a conscious decision on whether:

- We can (and therefore should) mitigate the Risk
- Pass the Risk back to the customer/client (i.e. exclude it), who only pays if it occurs
- Assume that the Risk will occur, and include it in the price the customer/client pays regardless (and Risk being unaffordable or uncompetitive)
- Run with the Risk and carry the can if we get unlucky

The major disadvantage in Monte Carlo Simulation for modelling Risk is that we forget that it is fundamentally flawed, and is inherently optimistically biased as it only models the Risks and Opportunities that we have thought about; it does not take account of Risks that we have not considered ... the Unknown Unknowns!

Now was that chapter as exciting and inspiring for you as it was for me? A case of real Monte Carlo Stimulation! *(You didn't think I'd let that one go unsaid, did you?)*

References

Augustine, NR (1997) *Augustine's Laws (6th Edition)*, Reston, American Institute of Aeronautics and Astronautics, Inc.

Book, SA (1997) 'Cost Risk Analysis: A tutorial', *Risk Management Symposium sponsored by USAF Space and Missile Systems Center and The Aerospace Institute*, Manhattan Beach, 2 June.

Book, SA (1999) 'Problems of correlation in the Probabilistic Approach to cost analysis', *Proceedings of the Fifth Annual U.S. Army Conference on Applied Statistics*, 19–21 October.

Iman, RL & Conover, WJ (1982) 'A distribution-free approach to inducing rank order correlation among input variables', *Communications in Statistics – Simulation and Computation*, Number 11, Volume 3: pp.311–334.

Metropolis, N & Ulam, S (1949) 'The Monte Carlo Method', *Journal of the American Statistical Association*, Volume 44, Number 247: pp.335–341.

Orwell, G (1945) *Animal Farm*, London, Secker and Warburg.

Peterson, I (1997) *The Jungles of Randomness: A Mathematical Safari*, New York, John Wiley & Sons, p.178.

RCPCH (2012a) *BOYS UK Growth Chart 2–18 Years*, London, Royal College of Paediatrics and Child Health.

RCPCH (2012b) *GIRLS UK Growth Chart 2–18 Years*, London, Royal College of Paediatrics and Child Health.

Rumsfeld, D (2002) *News Transcript: DoD News Briefing – February 12th*, Washington, US Department of Defense [online] Available from: http://archive.defense.gov/Transcripts/Transcript.aspx?TranscriptID=2636 [Accessed 13/01/2017].

Smart, C (2009) 'Correlating work breakdown structure elements', *National Estimator*, Society of Cost Estimating and Analysis, Spring, pp.8–10.

Stevenson, A & Waite, M (Eds) (2011) *Concise Oxford English Dictionary (12th Edition)*, Oxford, Oxford University Press.

Vose, D (2008) *Risk Analysis: A Quantitative Guide (3rd Edition)*, Chichester, Wiley.

4 | Risk, Opportunity and Uncertainty: A holistic perspective

Perhaps the two most widely used techniques to model Risks, Opportunities and Uncertainties are Monte Carlo Simulation (Chapter 3) and the Expected Value or Risk Factoring Technique that we will worry about (*literally*) in Chapter 5. There are other techniques we can use alongside Monte Carlo Simulation to give us a more holistic perspective of this troublesome topic.

4.1 Top-down Approach to Risk, Opportunity and Uncertainty

Whilst this section is fundamentally about Cost and Schedule Risk Analysis, some of the principles can be read across to other impacts of Risk, Opportunity and Uncertainty.

4.1.1 Top-down metrics

Most organisations will maintain records of past project performance against the agreed contractual price, schedule and performance criteria. If we can strip these back to the original cost estimate, baseline schedule and performance scope (i.e. remove any contingencies that were negotiated into the contractual agreements), then we can assess the level of change due to variability at a macro level. Whether or not we are able to categorise the changes as being due to unmitigated Risk, Risk Mitigation activities (baseline growth) or just worse than expected performance, it may be helpful, but not entirely necessary, as from a Top-down perspective we may be only interested in the overall impact. Every project will be different, but the level of control may be similar and the net impact may still be comparable.

A more helpful and easier piece of analysis may be the simple categorisation of projects or contracts along the lines of the natural Project Cycle in which the nature and level of Risks will change across the phases. For example:

Project Phase	Example Cause of Significant Change
• Concept Development	Requirements Change
• Design and Development	Technological Maturity
• Production	Supply Chain Stability
• In-service Support	Obsolescence
• Retirement and Disposal	Health and Safety Legislation

We might even be able to link the level of unmitigated Risk on a project or contract to the maturity of the underlying technology that was being used at the time of the project or contract. A commonly used categorisation for Technology Maturity is the NASA Technology Readiness Levels (Sadin et al, 1988) that we discussed in Volume I Chapter 6. Lower technology maturity could be indicative of elevated Risk and/or reduced performance due to unfamiliarity.

From this we may be able to identify ranges of cost growth or schedule slippage in relation to the original Baseline task in order to create generic Uplift Factors. For instance:

* Concept Development may carry an inherent Risk and Uncertainty of between 20% and 40% with the higher end being associated more with immature technology
* Production runs may carry an inherent Risk and Uncertainty exposure as low as 5% reducing to 2% with follow-on orders

Note: the percentages are for illustration purposes only and may differ across organisations and industries.

Caveat augur

We should resist the temptation to take the easy option and compare the outturn performance of any project with the agreed, contracted performance, which is likely to include a contingency for Risk Management. This will only tell us how we fared against the baseline estimate or plan.

If the contracted performance included a 10% contingency for Risk and Uncertainty, and we underperformed by some 5% then we should really have had a contingency of some 15.5% (i.e. 100% / 95%–100%).

4.1.2 Marching Army Technique: Cost-schedule related variability

The Marching Army Technique is a 'quick and dirty' technique. *(Visions of soldiers tramping through muddy terrain come to mind.)*

Most of us will probably agree that there is some correlation between cost and schedule variation; hence the expression 'Time is Money' (Franklin, 1748 in Labaree, 1961). If a project overruns its scheduled completion time, then we will need resources to complete the tasks after the scheduled completion time. More often than not this will imply additional cost as resources will need to be maintained on the project to complete the task. Even if we miss our schedule because the required resources were not available (implying that we have underspent) then it is possible that

> ### A word (or two) from the wise?
>
> *'Time is money.'*
> **Benjamin Franklin**
> American Statesman
> 'Advice to a Young Tradesman'
> 1746

the extension of the existing resource may attract higher rates of pay or vendor prices due to escalatory pressures.

In comparison, we can increase cost to avoid a schedule slip by loading additional resource on to a project.

The net result of all these dynamics is that we can assume a level of positive correlation between the two. Good historical records will help us here to understand the relationship in our organisations at different stages of the life cycle. Earned Value Management data can be particularly useful also in creating scatter diagrams of corresponding Cost Performance Indices (CPI) and Schedule Performance Indices (SPI). *(If we're lucky then we may get a good linear relationship for similar types of projects!)*

This technique assumes that we lead with a pessimistic view of the schedule; a Schedule Risk Analysis would be perfect but not a necessity.

* We are not necessarily looking at what is the 'doomsday' scenario of absolute worst case, but something more along the lines of the 'reasonable' values that we might expect when nothing seems to go right. It may help us to think of these in terms of the 80% or 90% Confidence Level rather than the Mode, Median or Mean of the schedule. In other words, 4 out of 5, or 9 out of 10 times, we would expect the schedule to be less than the pessimistic value we choose
* We then compare the ratio of the Pessimistic Schedule's overall time to completion with that of the Baseline Schedule (which presumably is based on the Mode, Median or better still, the Mean)
* We then increase the Cost Estimate by the same ratio on the basis that we have 'stood up' resources to meet the original schedule and these resources will be

required for longer at the corresponding levels. Alternatively, if we have already established a broad linear relationship between schedule slippage and cost growth then we should modify the ratio we apply accordingly

- We record our assumptions in the Basis of Estimate in the spirit of TRACEability (Transparent, Repeatable, Appropriate, Credible and Experientially-based)

Now, I can sense some rumblings amongst us that this is a very crude technique, and some of us may challenge whether it meets the 'Appropriate' criterion of TRACE, as normally any self-respecting organisation will try to manage its resource to avoid unnecessary cost, and if this is possible, then by all means we can exercise our judgement and modify the schedule slip to cost growth relationship accordingly. However, sometimes there is nowhere else to deploy the resources for short or long periods and the resource remains in place, even if the resource is booking to 'waiting time', it is still consuming cost … our 'Standing Army' continues to march. True, there may be a possibility of deferring the ramp-up of resource, and if this is the case then it can be taken into account.

We mustn't forget that the Marching Army Technique is a 'quick and dirty' technique. Just like the Grand Old Duke of York in the Nursery Rhyme, '*he marched them up to the top of the hill, and he marched them down again*' … in double-quick time through all the mud and the puddles.

Figure 4.1 shows our example pictorially, a 25% increase in schedule duration can result in 25% increase in cost. (Every point on the cost curve is shifted 25% to the right and 25% up.)

As there is an assumed (or empirically demonstrated where we have it) correlation between cost and schedule, then we can reverse the logic. Given a pessimistic view of the cost in relation to the Baseline Cost, we can factor the Baseline Schedule to determine an equivalent overall schedule timescale.

We can if we wish, break the schedule down into more discrete elements and apply the Marching Army Technique at a lower level of detail and then re-aggregate the cost profile accordingly.

We will revisit the use of this simplistic 'quick and dirty' technique as part of the overall decision-making process in Chapter 6.

4.1.3 Assumption Uplift Factors: Cost variability independent of schedule variability

Not all Cost Risks and Uncertainties have corresponding Schedule Risks and Uncertainties. For instance, Exchange Rate Uncertainties (often referred to as Exchange Rate Risks) will impact on cost to some degree depending on the level of foreign payments exposure we have, but there is no clear direct impact on schedule (except through

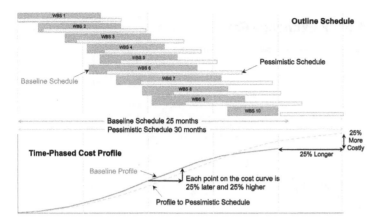

Figure 4.1 Example of Marching Army Technique

potential consequential cash flow issues of deferring a schedule). The level or rate of cost inflation or escalation would be another example:

For example, we may have included in our Basis of Estimate for the Baseline Estimate that we have assumed that inflation for the next 12 months will be 2.1% as advised by

> Note: Using Definition 3.1 for Uncertainty (Chapter 3), there will be an exchange rate, but we won't know the exact value.

some authoritative governmental or institutional body. However, this is still just an estimate by that authoritative body (*it hasn't happened yet!*), and there will be inherent uncertainty in the value they publish. (*They're just as capable of getting their estimates wrong as we are!*) We may want to take a more extreme pessimistic view and assume 5.5%, which would imply a 3.33% increase over the Baseline ($1.021 \times 1.033 = 1.055$).

Based on historical Risk and Opportunity Registers we may be able to identify minima and maxima variations in such variables over time. For this we may want to consider a reasonable range for an Uplift Factor that we might apply in comparison with the level we have assumed. We may want to consider using the various Moving Measures that we discussed in Volume III Chapter 3. (*I have assumed that you weren't so moved by it that you have chosen to try to forget it.*)

In Figure 4.2 we show an example of the Euro-GB Pound Exchange Rate history over a two-year period. Suppose that our Baseline assumptions is that based on the last six months the Exchange Rate will be around 1.39 € to £ 1.

However, a pessimistic view would be that for items purchased in Euros, the Exchange Rate could fall to around 1.20 € to £ 1. This might be considered by some to be an extreme value but if we were to look further back to 2007–2008, the Exchange Rate fell

Figure 4.2 Example of Historical Exchange Rate Data

from around 1.40 € per £ to 1.06 € per £ in just 15 months; in fact, it fell dramatically from 1.25 € per £ to its low of 1.06 € per £ in just two months. (*Now that's what makes for extreme estimating. Anyone not convinced that having to be pessimistic sometimes just goes with the job?*)

The reason for not taking the absolute extreme values will become clearer in the 'joined-up thinking' discussion we're going to have on this subject in Section 4.2.

Note also that the logic of this simplistic approach is not totally 'watertight' in the sense that some of these Uplift Factors may not be totally independent of schedule. For example, by slipping the schedule, we may incur different inflationary pressures than the baseline assumption; the main thing is not to 'double count' the effects.

We can use Uplift Factors in a more general way where we do not have a view of the pessimistic schedule. A checklist is often a useful *aide memoire* to consider areas of the project from which Risks or Opportunities might emanate; this could be a simple consideration of each functional area of the business (an Organisational Breakdown perspective) or one that is more of a Product-oriented Breakdown. Such checklists can be compiled from historical records of 'Lessons Learnt'. (*The cynics amongst us might say that if they keep recurring then we should refer to them as 'Lessons that should be learnt but somehow keep getting overlooked.'*)

4.1.4 Lateral Shift Factors: Schedule variability independent of cost variability

Just as there are some Uplift Factors we can apply to cost that are largely devoid of schedule issues, so too are there Lateral Shift Factors whereby we can vary the schedule to some degree without impacting on cost. For example, we may be able to place additional resource on tasks to reduce schedule task durations, but this does presuppose that the working shift arrangements are unchanged as that usually implies an impact on remuneration levels.

Lateral Shift Factors can be used to create the pessimistic view of the schedule used in the Marching Army Technique.

Note: It is common practice to depict cost on the vertical scale and schedule or time on the horizontal axis, hence the distinction between Uplift Factors and Lateral Shift Factors; in essence, they are the equivalent of each other.

4.1.5 An integrated Top-down Approach

The Marching Army, Uplift Factors and Lateral Shift Factors can all be used in combination with one another. We should, of course, make appropriate effort not to duplicate the effects. Due to the generally accepted concept that schedule and cost are partially correlated, the Marching Army Technique is probably the better technique to use first, supplemented by the Uplift or Lateral Shift Factors.

The overall evaluation using a Top-down Approach has all the hallmarks of being inherently pessimistic … if something can go wrong, it will go wrong.

4.2 Bridging into the unknown: Slipping and Sliding Technique

In Chapter 3 Section 3.4.1 we concluded that any Bottom-up Monte Carlo Simulation is inherently optimistic − a 'rose-tinted glasses' perspective. In Section 4.1.5 we have just concluded that the Top-down approach based on a combination of the Marching Army technique and Uplift and Lateral Factors is inherently pessimistic − a 'black tinted glasses' view of life. Perhaps then that reality lies somewhere between the two. The 'Slipping and Sliding Technique' (Jones & Berry, 2013) was developed at BAE Systems as part of an internal training course to allow delegates to perform a pseudo-calibration between the two approaches.

It begins with the Top-down Approach …

1. In order to perform a Risk Opportunity and Uncertainty Analysis we need to have an estimate for the Baseline task first. It doesn't have to be a Bottom-up estimate; a single-point deterministic Top-down estimate will suffice. However, in the spirit of TRACEability let's take the sum of the Most Likely values we had in Table 3.2 way back in Chapter 3 Section 3.1.6 (*practising what I preach about TRACEability … did you notice?*)

2. Next, if we now use the Marching Army Technique (Section 4.1.2) we can increase our Baseline cost estimate in proportion to the increase in schedule. Again, we will use the previous example which assumed 25% slippage

3. The final step is to make adjustments for non-schedule related cost variability. In our example from Section 4.1.3 we applied an Uplift Factor of 1.033 to take account of a pessimistic view of inflation of 5.5% instead of the Baseline assumption of 2.1%. In addition, suppose that 20% of our potential contract will be transacted in Euros. This was based on an Exchange Rate of 1.39 € to £ 1. As we saw in Section 4.1.3 a more pessimistic rate of 1.2 € to £ 1 is quite credible. This would give us an additional Uplift Factor of some 3.2%

This give us a 'final' pessimistic Top-down Estimate of £ 1,033 k, summarised in Table 4.1.

It may seem to be a very imprecise, 'rough and ready', 'quick and dirty' or whatever disparaging term we may want to use, but nevertheless it is still a valid perspective. The real value of an estimate generated in this way is that we can use it to compare with a Bottom-up Approach using Monte Carlo Simulation ...

For this we'll revisit our example using the Monte Carlo Simulation Output for the Baseline Uncertainty from our previous discussions in Chapter 3, Figure 3.46. (*Don't worry about flicking back to find it, we've reproduced it here in Figure 4.3.*)

1. We previously observed that the Mean Outcome was £ 809 k, but this is only equivalent to around 50% Confidence as the output distribution is very Norma-lesque. Suppose that we want to choose a slightly more conservative Confidence Level of some 60%, giving us a value of some £ 828 k
2. If we were to assume a simple 50:50 Risk Share based approximately on the 50% Confidence Level of Risks and Opportunities, this would give us a Risk Contingency of some £ 64 k (extracted from previous Figure 3.47)

Table 4.1 Top-Down Approach Using March Army Technique and Uplift Factors

Step	Description		Adjustments	Cost £k
1	Baseline Estimate: Sum of Most Likely Input Values			781
2	Schedule Slippage Penalty		25%	195
	Revised Estimate			976
3	Baseline Escalation assumed		2.1%	
	Pessimistic Escalation assumption		5.5%	
	Inflation Adjustment		1.033	33
	Proportion Transacted in Euros		20.0%	
	Baseline Exchange Rate	£1 = €	1.20	
	Pessimistic Exchange Rate assumption	£1 = €	1.39	
	Exchange Rate Adjustment to Baseline		3.2%	25
	Revised Estimate			1033

3. Adding this to our Baseline Estimate, we would get a value of some £ 892 k. This is equivalent to around the 56% Confidence Level on the overall Monte Carlo Simulation output for Risk, Opportunity and Uncertainty, as shown in Figure 4.3

When we make a simple comparison of this Bottom-up Estimate to the Top-down Estimate, there is a marked difference (£ 892 k vs. £ 1,033 k, or 15.9% difference). The question we should ask ourselves is 'are we really comparing "like for like"?' This is where the Slipping and Sliding Technique may help us make that 'fair' comparison. The procedure is relatively straightforward and is illustrated in Figure 4.4:

1. Draw an S-Curve of the Bottom-up Monte Carlo Confidence levels (the Cumulative Output Distribution) for the combined Risk, Opportunity and Baseline Uncertainty
2. Determine from the person who provided the pessimistic schedule and/or Uplift Factors, how pessimistic the Top-down Approach is. (This may be a qualitative perspective.) Let's say that in this example we are advised that it is probably around the 80% Confidence Level; in other words, there is a one in five chance that it could be even worse
3. Mark this point (£ 1,033 k at 80% Confidence) on the graph
4. Draw a second S-Curve that holds the 0% Confidence Level of the Bottom-up Curve fixed, but stretches the profile of the Bottom-up Curve to the right so that it passes through the Pessimistic Point determined at Steps 2 and 3. This is in effect

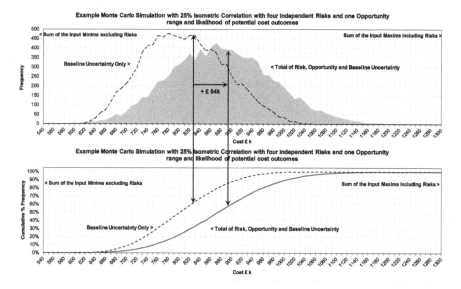

Figure 4.3 Choosing a Bottom-Up Estimate Using Monte Carlo Simulation Output

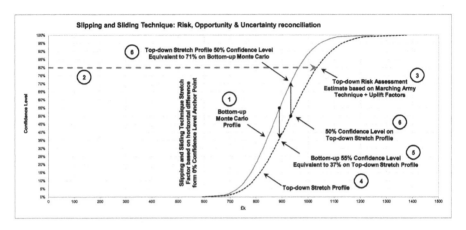

Figure 4.4 Slipping and Sliding Technique

Nominal Confidence Level:	0%	80%	Difference
Bottom-up Monte Carlo ROU:	£ 593 k	£ 963 k	£ 370 k
Top-down Marching Army + Uplift:	£ 593 k	£ 1033 k	£ 440 k
Slipping and Sliding (Stretch) Factor:			1.189

factoring the growth rate of the Bottom-up Curve by a fixed amount relative to 0% Confidence Level:

5. The Top-down pessimistic perspective makes some allowance for Unknown Unknowns and for more pessimistic values of Uncertainty in the Baseline and Risks. If we look at Bottom-up Estimate that we generated of £ 892 k we will see that this corresponds to the 37% Confidence Level of this Top-down pseudo profile

6. We can make the argument that if we feel that the Top-down perspective is credible, albeit pessimistic, then we should be looking at a Confidence Level on this factored curve of 50% as a minimum. In this case this would be equivalent to £ 933 k, or 71% Confidence Level on the Bottom-up Curve

7. We can make an informed judgement call on whether we proceed with this Top-down 50% Confidence Level. If we are of the view that perhaps this is a little too cautious, we could re-run the Technique with an increased Confidence Level for the Top-down Estimate of say 90%. In this way we can test the sensitivity of our assumption.

In this particular case the Bottom-up Curve and the more Pessimistic Top-down Stretched Curve appear to be relatively consistent with each other, but what if we had calculated different estimate values for our Top-down view? By our own conclusion the techniques used were a bit 'rough and ready'.

Suppose instead we had derived a Top-down Estimate of some £ 1,098 k based on a schedule slippage of some 33% using *Augustine's Law of Unmitigated Optimism*, (*Law Number XXIII*, 1997, p.152) that '*any task can be completed in only one-third more time than is currently estimated.*' (Let's leave the Uplift Factors as they are.) If we perform the Slipping and Sliding Technique now we get a Stretch Factor of:

$$\text{Stretch Factor} = (1098 - 593) / (963 - 593) = 1.365$$

Figure 4.5 illustrates that in this case we find that our Bottom-up Monte Carlo based estimate of £ 892 k is only equivalent to the 25% Confidence Level on our Top-down Stretch Profile. A minimum of 50% on the Top-down is equivalent to an 86% Confidence Level on the Bottom-up profile. From this we can conclude one or both of the following:

1. The Top-down Estimate using the Marching Army and Uplift Factors is overly pessimistic
2. The Bottom-up Analysis of Risks and Opportunities is understated. This may be due to immaturity in our understanding of the specific risks, or more worryingly, we are in 'risk denial'

In this latter case, if we choose to accept that the Top-down perspective is a valid one, then we can challenge the assumptions that have gone into the compilation of the Risk and Opportunity Register or Database.

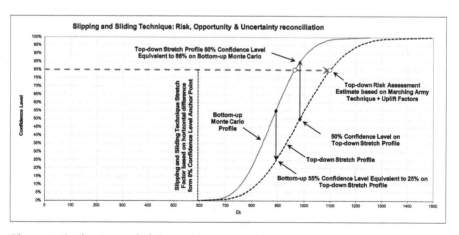

Figure 4.5 Slipping and Sliding Technique with Large Stretch Factor

To deal with the first case we can either re-run our Top-down Estimate with more conservative slip factors and Uplift Factors, or we can accept that our assumption of an 80% Confidence Level is understated. Suppose instead we recognise that it is overly pessimistic, and that the value is more comparable with a 90% Confidence Level, then we can re-run the technique to give us the 50% Confidence Level value of £ 941 k, equivalent to the 73% Confidence Level in the Bottom-up Monte Carlo profile, as shown in Figure 4.6.

Now let's consider the situation where we have Rambo (Morrell, 1972) as the testos-terone-fuelled Project Manager whose idea of a pessimistic schedule is a slippage of some 10% only. (*I can see from the fact that you are nodding sagely that you have met him or someone like him!*) In this case the Slipping and Sliding Technique will quickly highlight the anomaly that will arise, to the extent that we don't have to complete the technique. Figure 4.7 illustrates this.

The Marching Army Technique with a 10% Slippage plus the same Uplift Factors as we assumed previously (*Rambo has no input into these*) give us a Top-down Estimate of some £ 912 k. If this is assumed to be the equivalent of an 80% Confidence Level, then the comparable point on the Bottom-up Monte Carlo Simulation of the all-up Risk Opportunity and Uncertainty is £ 963 k.

We now have the bizarre situation where the Top-down pessimistic (black-tinted glasses) view of life is less than the Bottom-up optimistic (rose-tinted glasses) perspective! The Stretch Factor becomes a Compression Factor.

We can conclude that either or both of the following are true:

1. The Top-down perspective is naively optimistic (*which in the case of Rambo, it is what we expected*)

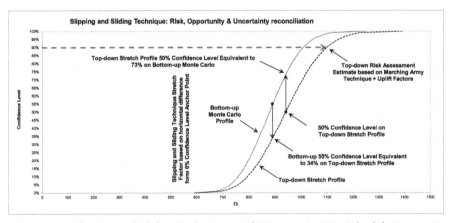

Figure 4.6 Slipping and Sliding Technique with Very pessimistic Schedule Assumption

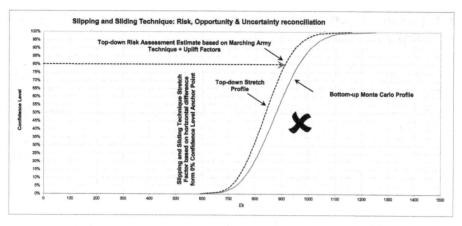

Figure 4.7 Slipping and Sliding Technique with a Nonsensical Stretch Factor of Less than One

2.   Our Bottom-up Monte Carlo perspective is overly pessimistic with inflated values or duplicated impacts

One benefit of the Slipping and Sliding Technique is that it considers more than one Approach. Management teams in general tend to place more trust (understandably to a degree) in the detailed Bottom-up Approach as it is more tangible, but they also like to make top-level comparisons with other projects that they can recall; this implicitly does that. However, the main benefit to the estimator (and therefore to the business) is that it allows us to make some provision for 'Unknown Unknowns' without artificially including a pseudo Risk with a random probability (thus breaking the principles of TRACEability). On the downside, the technique does pre-suppose that the profile of any alleged 'Unknown Unknowns' is similar to the 'Known Unknowns' of Table 3.23 in Section 3.4.1, which may or may not in fact be true.

We can extend the output of the Slipping and Sliding Technique as a guide to setting management reserve and Risk Contingency, if our internal processes allow these. Using our previous Monte Carlo Output data from Chapter 3 Section 3.3.2 and our Slipping and Sliding Technique from Figure 4.4 we can create optional models for budgeting purposes, slicing and dicing in alternative ways within a fixed (and hopefully agreed) total as illustrated in Table 4.2:

Table 4.2 Slipping and Sliding Technique as an Aid to Budgeting

Budget Element	Source	Option 1			Option 2		
		Confidence	Value	% Budget	Confidence	Value	% Budget
Estimate Recommendation	Top-down Profile	50%	933	100%	50%	933	100%
Internal Baseline Budgets	Monte Carlo Uncertainty	39%	781	84%	50%	806	86%
Risk Contingency	Monte Carlo Risk/Opp Only	50%	64	7%	64%	83	9%
Management Reserve	Balancing Figure	n/a	88	9%	n/a	44	5%
Estimate Recommendation	Bottom-up Monte Carlo (Total)	71%	933	100%	71%	933	100%

4.3 Using an Estimate Maturity Assessment as a guide to ROU maturity

For simplicity, in this section we shall just refer to Estimate Maturity Assessments (EMA) of Cost Estimates (Smith, 2013), but the same principles discussed will extend also to Schedule Maturity Assessments (SMA). See Volume I Chapter 3 for more information on both EMA and SMA.

There is no reason why we can't apply the principles of an Estimate Maturity Assessment to Risk, Opportunity and Uncertainty Modelling. In fact, the concept of an EMA is usually associated with the maturity of the Baseline task and its Basis of Estimate, and by implication, it covers its range of Uncertainty too. If we have developed the overall EMA score by a weighted average of the EMA scores for each constituent Task, then we will probably have used the Most Likely Values of each Task as the weightings. Strictly speaking, we should have used the Mean values, but as an EMA is not a precise measure then it is reasonably safe to assume that it is makes little tangible difference to the inherent message it is portraying.

Perhaps of greater significance is that an EMA 'roll-up' does not consider whether we have considered any specific or background Correlation between Baseline tasks. In the majority of cases, our confidence in the value of the correlation variable we have used is likely to be low (*assuming that we have applied it*) and if we are honest with ourselves we might rate this in the lower end of EMA Levels (1 to 4). If we haven't applied correlation in our Monte Carlo Simulation, then we should score it as zero; the more difficult question to answer is what weighting should we apply? Rather than over-engineer the weighting factor we apply, we can just assume a nominal 5% or 10% of the total. We mustn't lose sight of the fact that this is a qualitative indicator. By applying a nominal weighting of 5% or 10% for correlation, it will reduce higher EMA scores to a greater degree than lower end scores. (*This will stop us deluding ourselves that we can ever be perfect!*)

More importantly we need to consider our assessment of the Risk Contingency element of the estimate. We can apply the principles of EMA to this as well, scoring the Basis of Estimate for each Risk and Opportunity that we have defined. We must also consider our estimate of the Probability of Occurrence … how mature is that? Again, being somewhat brutal with ourselves, it is likely to score at the lower end of the EMA scale, especially if it has been generated by an Ethereal Approach with a Trusted Source (Expert Judgement) Method. If we have statistical evidence to back it up (e.g. the risk of component failure based on historical evidence), then it will score much higher. Again, the weighting we apply between the Probability of Occurrence variable and the Impact variable is a difficult decision, but the Probability of Occurrence has greater significance in the calculation of, and therefore our confidence in, the range of potential outcomes, and consequently we are better assuming a weighting contribution of the complement of its Probability as shown in Table 4.3 for our Cost Model example we have been using. Here we calculate the Adjusted EMA Score for each Risk or Opportunity by multiplying the EMA Impact Score by the Probability of Occurrence, and add this to the product of the EMA Probability of Occurrence Score and the Complement of the Probability (i.e. 100% – the Probability of Occurrence). In this example, the overall weighted EMA Risk and Opportunity Score of 3.72, reduces to 3.0 when we take into account our assessment on the Basis of Estimate for the Probability of Occurrence.

If we apply the same logic to Baseline Uncertainty, the weighting attributable to 100% Probability of Occurrence EMA Score is zero and the Adjusted EMA reduces to the same score as that for the EMA Impact value … in other words an ordinary EMA Score for an Uncertain Task. (*Don't you love it when it all hangs together logically?*)

However, the key element that is missing is the inevitable 'Unknown Unknowns'. These must implicitly score an EMA Level of 1, but the difficult decision again would be 'what weighting should we apply?' We could take a guess at a nominal value but there is probably little benefit in doing that; it would only reduce the EMA Score overall, even if we had made some allowance for 'Unknown Unknown' Risk arisings using the Slipping and Sliding Technique. We could artificially inflate the EMA Score for taking account of 'Unknown Unknown' arisings but this would be deluding ourselves and others. By implication they are covered in the lack of full maturity, i.e. a failure to score EMA Level 9.

We should really take a step back and assess our overall approach to developing an appropriate Risk Contingency and check for consistency. After all, we could score ourselves highly on our EMA Level for the Risk and Opportunity elements of our overall Estimate and then make an arbitrary decision on the level of Contingency Provision required, which is not in the spirit and intent of a mature, robust estimate.

> For instance, if we have used a Factored Value Technique (which we will discuss in Chapter 5) to assess the Risk Exposure and set the Risk Contingency, then this does not map obviously onto the EMA Framework. As the approach often hides or masks the true potential impact of low probability Risks, we should always view this as an

Table 4.3 Estimate Maturity Assessment for Defined Risks and Opportunities

ID	Cost Element	Distribution	Mode £k	Absolute Value Mode £k	Impact Weighting (% Sum of Abs Modes)	EMA of Risk / Opp Impact Value	Risk / Opp Probability of Occurrence	EMA of Risk / Opp Probability	Complement of Probability of Risk / Opp Occurrence	Adjusted Risk /Opp EMA Score
11	Risk 1	Uniform	60	60	20%	3	67%	4	33%	3.3
12	Risk 2	Beta	75	75	25%	6	10%	5	90%	5.1
13	Risk 3	Normal	50	50	17%	4	33%	3	67%	3.3
14	Risk 4	Triangular	60	60	20%	2	25%	1	75%	1.3
15	Opportunity 1	Triangular	-55	55	18%	3	20%	1	80%	1.4

| Total Value | | | 190 | 300 | 100% | | | | | |
| Weighted Average (% Sum of Modes) | | | | | | 3.72 | | 2.9 | | 3.0 |

immature approach when used in isolation. This would suggest an EMA rating in the lower level, e.g.

- Factored Most Likely   EMA 1 or 2
- Factored Mean   EMA 2 or 3
- Factored Maximum   EMA 3 or 4

Whereas a more mature approach would be to assess the Risk Exposure from different perspectives and then make an informed judgement of the appropriate Risk Contingency including some provision for those inevitable 'Unknown Unknowns'.

If we do conduct an EMA on the Risk and Opportunity Contingency element, then we might find it difficult to justify a higher rating for a Risk and Opportunity Contingency than for the Baseline task … *having a fat, dumb and happy Contingency does not mean that it is a mature estimate which implies a fair and honest assessment.* We can argue that there should be an implicit 'read-across' to what we would expect to see in an associated Risk Opportunity and Uncertainty Assessment, and this may lead us to choose one assessment method and technique over another, as indicated in Table 4.4.

4.4 Chapter review

The major disadvantage in Monte Carlo Simulation for modelling Risk and Opportunity, as we discussed in Chapter 3, is that we often forget that it is fundamentally flawed in being inherently optimistically biased, as it only models the Risks and Opportunities that we have thought about; it does not take account of all those valid, unsurfaced risks that we have not considered … the Unknown Unknowns!

We went on to discuss how we might counter the latter, and one way is to take a pessimistic approach to Risk, Opportunity and Uncertainty Evaluation. From a cost perspective, this could entail taking a pessimistic view of the schedule and increase cost pro rata to the slip in schedule. We called this the Marching Army Technique. To this we added Uplift Factors for a pessimistic view of non-schedule related costs such as annual escalation rates and exchange rates. Both these techniques assume that we do nothing to mitigate the potential cost increase.

Reality probably lies somewhere between the optimistic and pessimistic perspectives. We can only say 'probably' but there is still the chance that the Top-down Approach is too pessimistic or not pessimistic enough to cover the Unknown Unknowns to which we are unwittingly exposed. The Bottom-up Approach using Monte Carlo Simulation can also be naively understated, or overstated with duplicated risks, inflated probabilities of occurrence or impact values. In order to gain a measure of this and to help us make an appropriate decision we discussed the 'Slipping and Sliding Technique' as a form of pseudo-calibration between the two approaches.

Table 4.4 Estimate Maturity Assessment as a Guide to Risk, Opportunity & Uncertainty Technique

EMA Level	Baseline Estimate based on ...	Risk, Opportunity & Uncertainty Assessment characteristics
EMA9	Precise definition with recorded costs of the exact same nature to the Estimate required	Narrow Baseline Uncertainty that can be validated. Low number of risks with low Probability of Occurrence and Impact. Variability can be assessed by validated Top-down Metrics
EMA8	Precise definition with recorded costs for a well-defined similar task to the Estimate required	
EMA7	Precise definition with validated metrics for a similar task to the Estimate required	Suitable for Monte Carlo Simulation with an understood level of uncertainty in the Baseline Task. Risks and Opportunities have been defined but there is still the chance of Unknown Unknowns, which are not covered by the Bottom-up Approach. Top-down Marching Army Technique and/or Uplift Factors are used to supplement Monte Carlo
EMA6	Good definition with metrics for a defined task similar to the Estimate requred	
EMA5	Good definition with historical information comparison for a defined task similar to the Estimate required	
EMA4	Defined scope with good historical information comparison to the Estimate required	
EMA3	Defined scope with poor historical data comparison to the Estimate required	
EMA2	Poorly defined scope with poor historical data comparison to the Estimate required	Wide Baseline Uncertainty Range. Numerous Risks and Opportunities that are difficult to separate from the Baseline Scope. Potentially a large number of Unknown Unknowns. Top-down Marching Army Technique with Uplift Factors may be more appropriate than Monte Carlo Analysis
EMA1	Poorly defined scope with no historical data comparison to the Estimate required	

Finally, we discussed how we might use the Estimate Maturity Assessment Framework to calibrate the maturity of our Risk, Opportunity and Uncertainty Evaluation, and compare the appropriateness of our approach with the EMA score from our Baseline Estimate.

What we haven't discussed is the often-used technique of Risk Factoring and where it fits in the scheme of things. As we said in the opening paragraph, let's not worry about it now; we can do that in Chapter 5. Well, guess what, Chapter 5 is next.

References

Augustine, NR (1997) *Augustine's Laws (6th Edition)*, Reston, American Institute of Aeronautics and Astronautics, Inc.

Jones, AR & Berry, FJ (2013) 'An alternative approach to cost and schedule integration', SCAF Workshop on Cost and Schedule Integration, London, BAE Systems.

Labaree, LW (Ed.) (1961) *The Papers of Benjamin Franklin, Vol. 3, January 1, 1745, through June 30, 1750*, New Haven, Yale University Press, pp. 304–308.

Morrell, D (1972) *First Blood*, New York, M. Evans and Company.

Sadin, SR, Povinelli, FP & Rosen, R (1988) 'The NASA technology push towards future space mission systems', 39th International Astronautical Congress, Bangalore, India.

Smith, E (2013) 'Estimate Maturity Assessments', Association of Cost Engineers Conference, London, BAE Systems.

5 Factored Value Technique for Risks and Opportunities

In some organisations they take a 'swings and roundabouts' approach to Risk Contingency ... that everything will work itself out on average. A technique in popular use is that of the 'Factored Value' or 'Expected Value', in which we simply multiply the 'central value' of the Risk variables' 3-point range by the corresponding Probability of Occurrence. This includes Baseline Uncertainty with a probability of 100%. However, the popularity of a technique is probably more of an indicator of its simplicity than its appropriateness.

Depending on what the 'central value' represents, there are two ways we can do this ... the 'wrong way', and the 'slightly better way'. (*Did you notice that I didn't say 'right way'? Does that make it sound too much like I'm biased against this technique altogether?*)

Note that any factoring technique will inherently exclude any Unknown Unknowns and will therefore understate reality just as with Monte Carlo Simulation. The only possible exception to this might be the Factoring of the Maximum Values (*but even then we should expect some sleepless nights*).

5.1 The wrong way

As we commented previously, the 'central value' in a 3-Point Estimate is often the Most Likely (or Modal) value. Using the example that we created in Chapter 3, Table 3.19, and followed up in Chapter 4, Table 4.3, if we were to multiply the Mode of each Risk and Opportunity by its corresponding Probability of Occurrence, we would get a value of £ 68.2 k as illustrated in Table 5.1. To illustrate why this is an inherently misguided technique to use, let's consider the natural extension of this and apply the same logic to the Baseline Uncertainties. This then requires us to multiply the Most Likely Values by 100% and aggregate them (*in other words, just add them together*) as we have in the same table.

A word (or two) from the wise?

'Risk comes from not knowing what you're doing.'

Warren Buffett
American Business Executive
and Investor

Now some of us may already be shaking their heads saying, *'No, we shouldn't do that!'* (which is true) so why should we condone the practice for Risks and Opportunities?

However, setting that aside for a moment, let's follow the logic through, and compare the results with the Monte Carlo Output.

We might wonder if Warren Buffett was referring to the blind use of this technique when he expressed his views on the source of risk (Miles, 2004, p.85) ... *It would have been nice if he had been referring to this, but in reality he was talking in a more general sense of not investing in things we don't understand, if you read the full context. This can be read across here.*

We may recall that from our previous Monte Carlo Model (Chapter 3 Table 3.16), that the average or ExpectedValue for the Baseline tasks with 25% Isometric Correlation was £ 809.5 k, and for the four Risks and one Opportunity, it was £ 71.8 k (Figure 3.47 in Chapter 3 Section 3.3.2), or £ 880.3 k in total. As Table 5.1 shows that in this case

Table 5.1 Factored Most Likely Value Technique for Baseline and Risk Contingency

ID	Cost Element	Opt/Min £k	Mode £k	Pess/Max £k	Distribution	Probability	Factored Most Likely £k
1	Cost Element 1	50	80	110	Uniform	100%	80
2	Cost Element 2	80	101	150	Beta	100%	101
3	Cost Element 3	60	80	120	Beta	100%	80
4	Cost Element 4	20	50	80	Triangular	100%	50
5	Cost Element 5	30	50	100	Triangular	100%	50
6	Cost Element 6	75	100	150	Beta	100%	100
7	Cost Element 7	90	100	120	Beta	100%	100
8	Cost Element 8	70	80	110	Beta	100%	80
9	Cost Element 9	35	45	55	Uniform	100%	45
10	Cost Element 10	90	95	105	Triangular	100%	95
						Sum 1-10	781
11	Risk 1	50	60	70	Uniform	67%	40.2
12	Risk 2	65	75	95	Beta	10%	7.5
13	Risk 3	40	50	60	Normal	33%	16.5
14	Risk 4	55	60	80	Triangular	25%	15
15	Opportunity 1	-60	-55	-40	Triangular	20%	-11
						Sum 11-15	68.2
						Sum 1-15	849.2

the Factored Most Likely Technique under-
states the true Expected Value or Average of
the Baseline by £ 22.9 k or 3.5% and that the
Factored Most Likely Value of the assumed
Risk and Opportunity Distributions is also
understated by £ 3.6 k or some 5% in this
example.

> Expected Value = Value multi-
> plied by its probability of occur-
> ring (see Volume II Chapter 2)

> We may recall from Volume II Chapter 2 that for positively skewed distributions the
> Mode is less than the Median, i.e. 50% Confidence, which in turn is less than the
> Mean. This implies that any contingency based on the Factored Value of the Most
> Likely Values has less than 50% chance of occurring, or to put it another way we
> have a greater than 50% chance of overspending … *which doesn't sound like a sound
> basis for an important business decision, does it?*

Furthermore, as we have just discussed in Chapter 3 Section 3.3.3 a Contingency of £
64 k is not sufficient to cover Risk 2 should it occur. Nor is £ 68.2 k … unless we get
lucky and it comes in at the optimistic end!

Note: If we had had more distributions that were positively skewed rather than symmetrical, then this inadequacy would have been magnified.

Furthermore, if we had also included a Factored Most Likely Value for the Extreme
Risk we considered in Chapter 3 Section 3.3.6, then it would have only added an extra
£ 11 k (5% x £ 220 k). This also shows the futility of assigning Risk Contingency based
on the Factored Most Likely Value (*a regrettably not uncommon practice in some areas*) …
we either have £ 11 k more than we need if it does not occur, or we're between £ 189
k and £ 259 k short if it does occur (based on at a Minimum Value of £ 200 k and a
Maximum Value of £ 270 k).

Where has Rambo gone when you need him?

5.2 A slightly better way

If instead of taking the Factored Most Likely Value, what would happen if we were to
take the Factored Mean Value? Table 5.2 summarises the results and compares it with
the Factored Most Likely Value. (If the Mean values of each distribution are not readily
available, we could assume that they were all Triangular or Symmetrical Distributions
and take the average of the Minimum, Most Likely and Maximum Values; see Volume II
Chapters 3 and 4.)

In this particular case the Factored Mean for the Baseline Uncertainty is now in
line with our Monte Carlo Model Output, and the Factored Mean of the Risks and
Opportunity is some 3.7% higher than the Factored Most Likely. However, despite
being a better and more logical technique than factoring the Most Likely Value, the
value it gives is still less than the minimum for Risk 2 should it occur.

Table 5.2 Factored Mean Value Technique for Baseline and Risk Contingency

ID	Cost Element	Opt/Min £k	Mode £k	Pess/Max £k	Distribution	Probability	Mean £k	Factored Mean £k	Factored Maximum £k
1	Cost Element 1	50	80	110	Uniform	100%	80	80	
2	Cost Element 2	80	101	150	Beta	100%	105.67	105.67	
3	Cost Element 3	60	80	120	Beta	100%	82.50	82.5	
4	Cost Element 4	20	50	80	Triangular	100%	50	50	
5	Cost Element 5	30	50	100	Triangular	100%	60	60	
6	Cost Element 6	75	100	150	Beta	100%	105.00	105	
7	Cost Element 7	90	100	120	Beta	100%	101.67	101.67	
8	Cost Element 8	70	80	110	Beta	100%	82.50	82.5	
9	Cost Element 9	35	45	55	Uniform	100%	45	45	
10	Cost Element 10	90	95	105	Triangular	100%	96.67	96.67	
							Sum 1-10	809	
11	Risk 1	50	60	70	Uniform	67%	60	40.2	46.9
12	Risk 2	65	75	95	Beta	10%	76.67	7.67	9.5
13	Risk 3	40	50	60	Normal	33%	50	16.5	19.8
14	Risk 4	55	60	80	Triangular	25%	65	16.25	20
15	Opportunity 1	-60	-55	-40	Triangular	20%	-51.67	-10.33	-8
							Sum 11-15	70.28	88.20
							Sum 1-15	879.28	

Note that if we were to take the Factored Maximum Values of the Risks and Opportunity in this case, then we would get a value that would indeed cover off this largest Impact Risk if it were to occur. However, we cannot generalise and say that would always be the case, but it may be an indication that the better value for unmitigated Risk Contingency is at least the Factored Input Means and potentially nearer to the Factored Input Maxima Values … which might make one or two of us a little uncomfortable, and potentially it may make us uncompetitive. If nothing else, it gives us a strong indication of the true size of the problem to be managed. Figure 5.1 illustrates the range of values in comparison with the Monte Carlo Output. In this case the Confidence Level of the Factored Input Means is around 60%.

Even then, for a High Impact, Low Probability Risk, the contribution to Risk Contingency made by Factoring even a maximum value, fails the Credibility element of our TRACEability objective

5.3 The best way

If in doubt, don't do it!

Don't Factor Risk values by their Probabilities of Occurrence except as a means of validating a more analytical review of the data, such as the approach suggested in Chapter 3 Section 3.4.

If we have no alternative, i.e. we do not have access to suitable Monte Carlo Simulation software, then we should consider using the Factored Maxima of the Risks (not Opportunities) as this will at least provide some provision for a higher Confidence

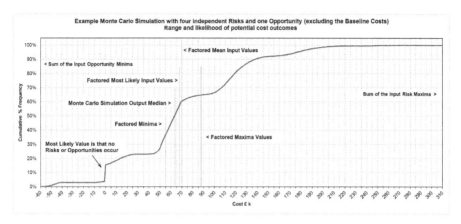

Figure 5.1 Example Monte Carlo Simulation with Four Independent Risks and One Opportunity (Excluding the Baseline Costs) - Range and Likelihood of Potential Cost Outcomes

Level that may help to alleviate some the pain we will feel in the event of the inevitable Unknown Unknowns. To this we need to add the Means of the Baseline Uncertainty (not the Modes or the Medians!).

In the case of this example from Table 5.2 we would get £ 97.85 k + £ 809 k or £ 906.85 k ... which is still less than the £ 933 k that we compiled using the Slipping and Sliding Technique supported by two different approaches, and it is equivalent to around 61% Confidence Level on the Bottom-up Monte Carlo Simulation, or the 41% Confidence Level on our pseudo Top-down profile.

Factoring .... use it only as a last resort, but if you can avoid it, then avoid it by all means!

5.4 Chapter review

We recognised that some organisations like to use the very simplistic Factoring Technique for developing a view of Risk Contingency. Many simply take the sum of the product of each Risk's Most Likely Value multiplied by its Probability of Occurrence. This will understate the true Expected Value of the identified Risks; we really should use their Mean values not their Most Likely ones. Factoring the maximum values is equivalent to taking a higher Confidence Level from a Monte Carlo Simulation, but may still be less than the value indicated by the Slipping and Sliding Technique in Chapter 4. Furthermore, whichever factoring technique we use it will suffer even more acutely from the same problem as Bottom-up Monte Carlo Simulation ... in that it does not include any provision whatsoever for Unknown Unknowns. At least Monte Carlo allows us to err on the cautious side with a higher Confidence Level.

Does it pass the TRACEability test? Hardly! It may be a technique that is Transparent, certainly Repeatable, and is often used by experienced people, but where it falls down is on whether it is Appropriate, and it often fails the Credibility test in that it gives next to nothing as a contribution towards Low Probability, High Impact Risks.

Reference

Miles, RP (2004) *Warren Buffett Wealth: Principles and Practical Methods Used by the World's Greatest Investor*, New York, John Wiley & Sons, p.85

6 Introduction to Critical Path and Schedule Risk Analysis

Recognising the need to plan is the first step towards arriving at a successful outcome. The reverse is certainly true, as the popular unverified quotation states: if we don't plan, we shouldn't expect to be successful ... but we might just get lucky! (*Hmm, try telling your director that and see how long you get to stay in the same room!*)

Every estimator should be able to plan in support of a robust Estimating Process (see Volume I Chapter 2). If they are cost estimators, they also need to be able to understand the implications of a plan or schedule in terms of the interaction between cost and schedule.

If everything in life was sequential, the life of the planner, scheduler and estimator would

be so much simpler, but everyone else would have to learn to be patient as many things would take a lot longer to achieve than they do now, when inevitably, some tasks are performed in parallel. So, how much work can be performed in parallel? This is where Critical Path Analysis can come to our aid.

6.1 What is Critical Path Analysis?

To speed things up, we can (and do) allow some tasks or activities to occur in parallel, but there is normally a physical limit or constraint in terms of a natural sequence of pre-requisite tasks or activities. For instance, we cannot derive the numerical value of an estimate in parallel to the agreement of the assumptions which underpin it; the latter is a pre-requisite of the former. The overall time from start to finish is determining by this natural sequence and the task/activity durations.

We can draw these tasks/activities and the inherent dependencies as a logic network diagram. A search of the internet will show us that there are many different styles in use. Some are better than others in conveying the essential information; the key thing is to choose a style that works best for us and our individual organisations. The styles really fall into two camps:

- **Activity on Node:** Each task or activity is drawn on node represented by a geometric shape (e.g. circle or rectangle). They are joined by lines indicating the direction of logical flow to our next task or activity and are labelled or annotated accordingly
- **Activity on Arrow:** Each task or activity begins and ends at a node. The nodes represent the logical points by which parallel tasks/activities must 'come together' or can divide. The arrows or boxes between the nodes represent the tasks/activities and are labelled or annotated accordingly

The essential information we must input to the network is:

- Any required preceding tasks or activities
- Any required succeeding tasks or activities
- Each task or activity's expected duration

From the network, we can derive the earliest and latest start and finish times for each task or activity, and any slack or float time that can be used to vary start or finish times of individual tasks or activities, should the need arise. It also allows us to identify the Critical Path, i.e. that route through the network where no float is available if we want to achieve the earliest start to finish time for the project overall.

Definition 6.1 Critical Path

The Critical Path at a point in time depicts the string of dependent activities or tasks in a schedule for which there is no float or queuing time. As such the length of the Critical Path represents the quickest time that the schedule can be expected to be completed based on the current assumed activity durations and associated circumstances.

The question now arises on whether the earliest start time of the first activity is at time zero, or time 1 (*or any other time for that matter*). Furthermore, we might also ask ourselves whether we are assuming that each activity ends in the middle of the time period, or at the end, and whether the next activity starts immediately or in the next time period. For example, consider two activities with durations of four and five time periods; Table 6.1 summarises the two options available to us. It doesn't matter which option we choose so long as we are consistent. We can always work with the simplest option first (Option 0, the null option) and convert to the other option later. We will be adopting Option 0 in the examples that follow.

Let's look at an example. In Figure 6.1 we have shown the Network Diagram for a Plan to develop a Time-phased Estimate including an appropriate level of Risk and Uncertainty Contingency using an 'Activity on Node' format, and using the start/finish times at the mid-point of the time intervals (Option 0).

We will notice that there is a table of data underpinning each activity (*we will get back to the calculations shortly ... I know, I can hardly wait myself*):

Top-left corner:	The earliest start time (ES) for that activity, assuming that all predecessor activities have been completed to plan and that the current activity starts as soon as the last predecessor has completed
Top middle:	The planned duration time for the activity
Top-right corner:	The earliest finish time (EF) for that activity, assuming that all predecessor activities have been completed to plan and that the current activity continues uninterrupted
Bottom-left corner:	The latest start time (LS) for that activity such that no subsequent activity on the Critical Path is impeded and that the activity requires its planned duration
Bottom middle:	The activity float or slack time representing the amount of time an activity can move to the right without impacting on any subsequent activity
Bottom-right corner:	The latest finish time (LF) for that activity such that no subsequent task impedes the Critical Path

Most scheduling applications will allow us to calculate the Critical Path as standard functionality, but if we don't have access to such a tool, we can do relatively simple networks using Microsoft Excel to calculate the Critical Path and the table of data described above.

Table 6.1 Activity Start and End Date Options

	Activity 1		*Activity 2*	
Duration	4		5	
	Start Time	Finish Time	Start Time	Finish Time
Option 0	0	4	4	9
Option 1	1	4	5	9

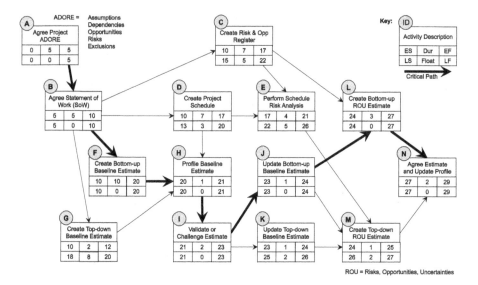

Figure 6.1 Critical Path for an Example Estimating Process

6.2 Finding a Critical Path using Binary Activity Paths in Microsoft Excel

… with a little bit of help from Microsoft PowerPoint (or even a blank sheet of paper and a pen).

We can generate the Critical Path with the following procedure:

1. List all the activities and their expected or planned durations
2. Draw the network relationships showing which activity is dependent on the completion of others (*just like we did in Figure 6.1 … but without any calculations or data*). It's a good idea also to give each activity a simple unique ID such as a letter in such a way that each path follows in an alphabetic order. (*It doesn't matter if there are 'jumps' to the next letter, but it mustn't go back on itself.*)
3. Using the simple ID, list ALL unique paths through the network as a string of ID characters, e.g. ABCLN, ABCEMN, ABDEMN etc. from Figure 6.1
4. Create a Binary Activity Matrix (as shown in Table 6.2) so that the activity is represented as being present in the path by a 1, and absent by a 0 or blank (here we have used a blank). If we create this table manually there is always a risk of an error. We can help to mitigate against this by adding 'Checksums' to compare the Length of the Activity Path in terms of the number of characters using **LEN(*Path*)** with the sum of the Binary Values we have input. Here **Path** is the Cell containing the

Activity Path, e.g. ABCLN. (This is not a perfect validation as we could still enter it in the wrong space.)

Alternatively, for the more adventurous amongst us, we could generate it 'automatically' in Microsoft Excel using **IF(IFERROR(FIND(*ID,Path*,1),0)>0,1,"")** where **ID** is the Activity Character Cell recommended in Step 2 and **Path** is the Activity Path Cell from Step 3

5. We can find the minimum length of each path by aggregating the durations of each activity in the path. The easiest way of doing this in Microsoft Excel is to use the **SUMPRODUCT(*range1, range2*)** function where one range is fixed as the range of activity durations, just under the activity ID in our example; the other range is the binary sequence for the path in question

6. The Critical Path is that path which has the maximum activity duration (i.e. the maximum of the minima, *if that doesn't sound like too much of an oxymoron!*). In this example it is Activity Path 7

7. As a by-product of this process, it allows us to calculate the total float or slack for each Activity Path in a network by calculating the difference between the Critical Path's duration and the individual Activity Path Durations. (Again we have shown this in Table 6.2)

Table 6.2 Binary Activity Matrix Representing Each Activity Path Through the Network

Column key (A–N): A = Agree Project ADORE; B = Agree SoW; C = Create Risk Opp Register; D = Create Project Schedule; E = Perform Schedule Risk Analysis; F = Create Bottom-up Baseline Estimate; G = Create Top-down Baseline Estimate; H = Profile Estimate; I = Perform Validation or Challenge; J = Update Bottom-up Baseline Estimate; K = Update Top-down Baseline Estimate; L = Create Bottom-up ROU Estimate; M = Create Top-down ROU Estimate; N = Complete Estimate Recommendation

ID	Activity Path	A	B	C	D	E	F	G	H	I	J	K	L	M	N	Total Path Duration	Total Path Float (Max - Path Duration)	Checksum Path Length	Checksum Sum Binaries
	Duration	5	5	7	7	4	10	2	1	2	1	1	3	1	2				
1	ABCLN	1	1	1									1		1	22	7	5	5
2	ABDEMN	1	1		1	1								1	1	24	5	6	6
3	ABCEMN	1	1	1		1								1	1	24	5	6	6
4	ABDHIJLN	1	1		1				1	1	1		1		1	26	3	8	8
5	ABDHIJMN	1	1		1				1	1	1			1	1	24	5	8	8
6	ABDHIKMN	1	1		1				1	1		1		1	1	24	5	8	8
7	ABFHIJLN	1	1				1		1	1	1		1		1	29	0 < Critical Path	8	8
8	ABFHIJMN	1	1				1		1	1	1			1	1	27	2	8	8
9	ABFHIKMN	1	1				1		1	1		1		1	1	27	2	8	8
10	ABGHIJLN	1	1					1	1	1	1		1		1	21	8	8	8
11	ABGHIJMN	1	1					1	1	1	1			1	1	19	10	8	8
12	ABGHIKMN	1	1					1	1	1		1		1	1	19	10	8	8

Maximum Path Duration: 29

8. We can now calculate the earliest start and finish times for our network in relation to our Critical Path. Table 6.3 illustrates how we can do this by creating a mirror image table of our Binary Matrix. Each value in the table represents the completion time of that activity based on all prior activities in that path being completed. We can do this in Microsoft Excel with two calculations:

   • For the first activity (A in this case), we assume that it starts at time 0 and finishes when the time matches its planned or expected duration
   • For all other a, to calculate its earliest finish time we must add its expected duration to the maximum of earliest finish times for the previous activity in its path (ignoring all other activities that do not occur in its path). In Microsoft Excel we can do this with a Condition formula that refers to the Binary Matrix (explained here in words):

Table 6.3 Calculation of Earliest Finish and Start Dates in an Activity Network

		Binary Path Matrix													
ID	ID	A	B	C	D	E	F	G	H	I	J	K	L	M	N
	Duration	5	5	7	7	4	10	2	1	2	1	1	3	1	2
ID	Activity Path														
1	ABCLN	1	1	1									1		1
2	ABDEMN	1	1		1	1								1	1
3	ABCEMN	1	1	1		1								1	1
4	ABDHIJLN	1	1		1				1	1	1		1		1
5	ABDHIJMN	1	1		1				1	1	1			1	1
6	ABDHIKMN	1	1		1				1	1		1		1	1
7	ABFHIJLN	1	1				1		1	1	1		1		1
8	ABFHIJMN	1	1				1		1	1	1			1	1
9	ABFHIKMN	1	1				1		1	1		1		1	1
10	ABGHIJLN	1	1					1	1	1	1		1		1
11	ABGHIJMN	1	1					1	1	1	1			1	1
12	ABGHIKMN	1	1					1	1	1		1		1	1

		Earliest Finish Dates for Path Working Forwards >>>														
ID	ID	A	B	C	D	E	F	G	H	I	J	K	L	M	N	
	Duration	5	5	7	7	4	10	2	1	2	1	1	3	1	2	
ID	Activity Path															
1	ABCLN	5	10	17									20		22	
2	ABDEMN	5	10		17	21								22	24	
3	ABCEMN	5	10	17		21								22	24	
4	ABDHIJLN	5	10		17				18	20	21		24		26	
5	ABDHIJMN	5	10		17				18	20	21			22	24	
6	ABDHIKMN	5	10		17				18	20		21		22	24	
7	ABFHIJLN	5	10				20		21	23	24		27		29	
8	ABFHIJMN	5	10				20		21	23	24			25	27	
9	ABFHIKMN	5	10				20		21	23		24		25	27	
10	ABGHIJLN	5	10					12	13	15	16		19		21	
11	ABGHIJMN	5	10					12	13	15	16			17	19	
12	ABGHIKMN	5	10					12	13	15		16		17	19	
	Earliest Finish (EF)	5	10	17	17	21	20	12	21	23	24	24	27	25	29	Individual Activity EF Maxima
	Earliest Start (ES)	0	5	10	10	17	10	10	20	21	23	23	24	24	27	Max EF - Duration

IF(*logical_test, value_if_true, value_if_false*)

Where:

logical_test: Checks whether Activity's corresponding Binary Matrix Cell is blank or zero,

value_if_true: Enters "" as the cell value

value_if_false: Finds the maximum value of all the cells to the left of the current cell on the current Activity Path (row) and adds the product of the current cell's corresponding Binary Matrix Cell Value (0 or 1) multiplied by the Activity Duration. For example, Activity Paths 7, 8 and 9 all add 1x10 to the finish date of Activity B to get the finish date for Activity F

9. The maximum value of each column represents the earliest finish time of each activity (EF)
10. We can determine the earliest start time (ES) of each activity by subtracting the activity duration

6.3 Using Binary Paths to find the latest start and finish times, and float

The procedure for finding the latest start and finish times for each activity is very similar, except that in this case we will find the latest start first, and that we will work backwards from the last task.

Steps 1–7 are identical to those in Section 6.2:

8. We can calculate the latest start and finish times for our network in relation to our Critical Path. Table 6.4 illustrates how we can do this by creating a mirror image table to our Binary Matrix. Each value in the table represents the completion time of that activity based on all subsequent activities in that path being completed. We can do this in Microsoft Excel with two calculations:

- For the last activity (N in this case), we assume that it starts at the time indicated by the end of the Critical Path (Time = 29 in this case) less the expected or planned Activity Duration
- For all other activities, to calculate the latest start time we must subtract its expected duration from the minimum of latest start times for the next activity in its path (ignoring all other activities that do not occur in its path). In Microsoft Excel we can do this with a Condition formula that refers to the Binary Matrix (explained here in words):

IF(*logical_test, value_if_true, value_if_false*)

Where:

Table 6.4 Calculation of Latest Finish and Start Dates in an Activity Network

ID		A	B	C	D	E	F	G	H	I	J	K	L	M	N
	Binary Path Matrix														
	Duration	5	5	7	7	4	10	2	1	2	1	1	3	1	2
ID	Activity Path														
1	ABCLN	1	1	1									1		1
2	ABDEMN	1	1		1	1								1	1
3	ABCEMN	1	1	1		1								1	1
4	ABDHIJLN	1	1		1			1	1	1		1			1
5	ABDHIJMN	1	1		1			1	1	1				1	1
6	ABDHIKMN	1	1		1			1	1		1			1	1
7	ABFHIJLN	1	1				1		1	1	1		1		1
8	ABFHIJMN	1	1				1		1	1	1			1	1
9	ABFHIKMN	1	1				1		1	1		1		1	1
10	ABGHIJLN	1	1					1	1	1	1		1		1
11	ABGHIJMN	1	1					1	1	1	1			1	1
12	ABGHIKMN	1	1					1	1	1		1		1	1

ID		A	B	C	D	E	F	G	H	I	J	K	L	M	N
	Binary Path Matrix														
	Duration	5	5	7	7	4	10	2	1	2	1	1	3	1	2
ID	Activity Path														
1	ABCLN	7	12	17									24		27
2	ABDEMN	5	10		15	22								26	27
3	ABCEMN	5	10	15		22								26	27
4	ABDHIJLN	3	8		13			20	21	23		24			27
5	ABDHIJMN	5	10		15			22	23	25				26	27
6	ABDHIKMN	5	10		15			22	23		25			26	27
7	ABFHIJLN	0	5				10		20	21	23		24		27
8	ABFHIJMN	2	7				12		22	23	25			26	27
9	ABFHIKMN	2	7				12		22	23		25		26	27
10	ABGHIJLN	8	13					18	20	21	23		24		27
11	ABGHIJMN	10	15					20	22	23	25			26	27
12	ABGHIKMN	10	15					20	22	23		25		26	27

Row 7 note: 29 | < Critical Path (appears after the 27 value in the ABFHIJLN row)

		A	B	C	D	E	F	G	H	I	J	K	L	M	N	
Latest Start (LS)		0	5	15	13	22	10	18	20	21	23	25	24	26	27	Individual Activity LS Minima
Latest Finish (LF)		5	10	22	20	26	20	20	21	23	24	26	27	27	29	Min LS + Duration

logical_test:	Checks whether the Activity's corresponding Binary Matrix Cell is blank or zero,
value_if_true:	Enters "" as the cell value
value_if_false:	Finds the minimum value of all the cells to the right of the current cell on the current Activity Path and subtracts the product of the current cell's corresponding Binary Matrix Value multiplied by the Activity Duration

9. The minimum value of each column represents the latest start time of each Activity (LS)
10. We can determine the latest finish time (LF) of each activity by adding the activity duration

We can determine the total float or slack for each activity by calculating the difference between the latest and earliest finish times (or start times) from Tables 6.3 and 6.4 as illustrated in Table 6.5. (*The US Navy who first developed the concept of Critical Path Analysis preferred the term 'Slack', presumably because they were uncomfortable with the concept of zero float!*)

6.4 Using a Critical Path to Manage Cost and Schedule

Cynics of Critical Path Analysis may claim that there is a risk of an 'over-focus' on the Critical Path activities which can lead to cost overruns on non-Critical Path activities. If we take that concern on board, we could use the concept of float to determine our exposure to any such cost overrun attributable in this way to non-Critical Path activities. In Table 6.6 we have allowed all the non-Critical Path activities in our example to commence at the earliest start time and to consume all the float to finish at the latest time without impacting on the Critical Path. However, we have allowed these activities to continue consuming resources unabated using the 'Standing or Marching Army' principle that we discussed in Chapter 4. This shows that we are exposed to a 34% overspend without impacting on the Critical Path.

On the other hand, supporters of Critical Path Analysis can point to the opportunity to use activity float to smooth resource loading across a project, especially where limited resource is required on parallel tasks.

Let's consider a very simplistic profiling of estimating resource over our Estimating Plan, assuming a uniform distribution of the resource hours over each activity (*this may be too simplistic for some 'macro' activities*). Figure 6.2 aggregates the total resource hours

Table 6.5 Calculation of Activity Float in an Activity Network

	Activity													
ID	A	B	C	D	E	F	G	H	I	J	K	L	M	N
Duration	5	5	7	7	4	10	2	1	2	1	1	3	1	2
Latest Start (LS)	0	5	15	13	22	10	18	20	21	23	25	24	26	27
Latest Finish (LF)	5	10	22	20	26	20	20	21	23	24	26	27	27	29
Earliest Start (ES)	0	5	10	10	17	10	10	20	21	23	23	24	24	27
Earliest Finish (EF)	5	10	17	17	21	20	12	21	23	24	24	27	25	29
Activity Float	0	0	5	3	5	0	8	0	0	0	2	0	2	0

Table 6.6 Calculation of Activity Float in an Activity Network

	Activity														Total Hours
	A	B	C	D	E	F	G	H	I	J	K	L	M	N	
Activity Duration	5	5	7	7	4	10	2	1	2	1	1	3	1	2	
Activity Float	0	0	5	3	5	0	8	0	0	0	2	0	2	0	
Maximum Slip Exposure	0%	0%	71%	43%	125%	0%	400%	0%	0%	0%	200%	0%	200%	0%	
Activity Hours	20	30	21	21	8	60	6	2	6	3	2	9	2	6	196
Overspend Exposure	0	0	15	9	10	0	24	0	0	0	4	0	4	0	66
															34%

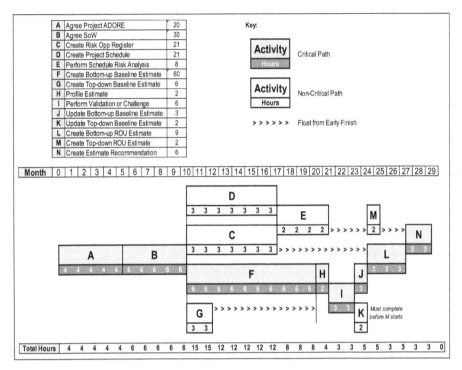

Figure 6.2 Resource Profiling (Unsmoothed) Using Earliest Start and Finish Dates Around Critical Path

using the earliest start and finish times for each activity; the overall resource varies from time period to time period. Figure 6.3 shows a potential smoothing of the hours can be achieved by allowing some of the non–Critical Path activities to start later.

If we want to smooth the schedule resourcing any further, then we will need to consider stretching or compressing individual activity durations. Sometimes this action may be forced on us to recover a plan … but beware, we cannot always assume that activities will cost the same as the 'ideal' plan when we do so.

Increasing resource on an activity may increase inefficiencies in people obstructing access to the work space inadvertently, or may increase the need for communication and handover arrangements between team members. This will especially be true if shortening an activity requires a change in shift arrangements, e.g. one shift to two shifts.

Similarly, removing resource from an activity will increase its duration but the cost may not reduce in inverse proportion as we might expect due to 'Parkinson's Law' (Parkinson, 1955) where '*work expands so as to fill the time available for its completion.*' Augustine (1997, p.158) referred this duality of cost behaviour as the '*Law of Economic Unipolarity*'.

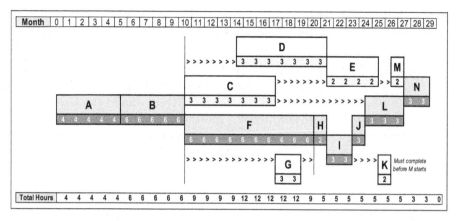

Figure 6.3 Resource Profile Smoothing Using Activity Float Around the Critical Path

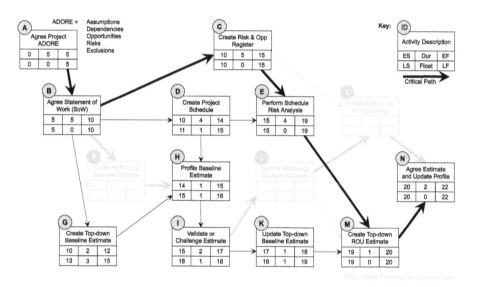

Figure 6.4 Critical Path Accelerated by Removing Activities

In some instances, we may find that the Critical Path is 'not good enough' (i.e. too long) to meet the business requirement. We may have to resort to increasing resources to accelerate the Critical Path activities need and take the consequential hit on the cost of doing so. However, we may be able to review the solution we have proposed and plan to remove activities from the Critical Path. For example, in the case of our plan to generate a time-phased estimate, we may conclude that we can dispense with the need for a Bottom-up Baseline Estimate and just rely on a Top-down Estimate. This in effect may be sacrificing quality for schedule (and cost as a by-product). Figure 6.4 shows such an accelerated plan with a revised Critical Path.

Here we have removed the three activities that relate to creating Bottom-up Estimates for the Baseline condition and for Risk, Opportunity and Uncertainty. We have also reduced the planned activity duration for Scheduling and creating the Risk and Opportunity Register. This leaves us with only four potential activity paths: ABCEMN, ABDEMN, ABDHIKMN and ABGHIKMN, of which the first becomes the Critical Path with a duration of 22 days.

6.5 Modelling variable Critical Paths using Monte Carlo Simulation

Until now we have only considered Critical Paths with fixed duration activities … *until we started messing around thinking about changing them as a conscious act.* However, in many cases we may not know the exact duration an activity will take as there will be Uncertainty around the potential durations and the estimates of the resource times required. In these cases it would be more appropriate to model the schedule with Uncertainty ranges around our activity durations. This leads us to Monte Carlo Simulation and the generation of variable Critical Paths (*literally*).

If we accept the principle that we discussed in Chapters 3 and 4, there will be an Uncertainty range around the duration of most, if not all, the activities in the network. This then raises the distinct possibility that the Critical Path may not be totally definitive, and that we might only be able to express that a particular activity path is the Critical Path with a specified level of confidence. This may lead us to having to manage activities that are 'off' the supposed Critical Path more closely than if they were on the Critical Path; in other words, give them similar priority or due diligence as we give to definitive Critical Path activities.

Consider Activity Path 4 (ABDHIJLN) from our earlier Table 6.3 and compare it with our Critical Path or Activity Path 7 (ABFHIJLN). There is only one activity difference between them; the former includes Activity D whereas the latter includes Activity F. If our Critical Path Activity F (Create Bottom-up Baseline Estimate) takes two days less than anticipated, but Activity D (Create the Project Schedule) takes two days more to complete, then Activity Path 4 becomes the Critical Path, not Activity Path 7.

We can extend this Uncertainty to other activity durations (Table 6.7).

If we were to run a Monte Carlo Simulation using these Discrete Uniform Distribution Uncertainty Ranges in Table 6.7, we would find that Activity Path 7 is only the Critical Path some 85% of the time. Activity Paths 2–4 are also contenders for the Critical Path as summarised in Table 6.8. This tells us that either we can't add up, or that 14% of the time

Table 6.7 Uncertainty Ranges Around Activity Durations

ID	Activity	Min	Mode	Max
A	Agree Project ADORE	4	5	6
B	Agree SoW	4	5	6
C	Create Risk Opp Register	5	7	9
D	Create Project Schedule	5	7	9
E	Perform Schedule Risk Analysis	3	4	5
F	Create Bottom-up Baseline Estimate	6	10	14
G	Create Top-down Baseline Estimate	1	2	3
H	Profile Estimate	1	1	1
I	Perform Validation or Challenge	1	2	3
J	Update Bottom-up Baseline Estimate	1	1	1
K	Update Top-down Baseline Estimate	1	1	1
L	Create Bottom-up ROU Estimate	2	3	4
M	Create Top-down ROU Estimate	1	1	1
N	Create Estimate Recommendation	1	2	3

Table 6.8 Critical Path Uncertainty

Activity Path		Occurrences as Critical Path	Comment
7	ABFHIJLN	85%	Includes instances
4	ABDHIJLN	20%	of 2 or more joint
3	ABCEMN	6%	Critical Paths
2	ABDEMN	3%	
	Total	114%	

we have instances where more than one path is the Critical one. Looking at the paths in detail the only activities that are not covered are G and K, which are the creation and update of the Top-down Estimate.

In Figure 6.5, we can show that if Activity Path 7 exceeds its Most Likely Duration of 29 Days, then it is highly probable that it is the Critical Path, and that instances of it not being the Critical Path are associated with the more optimistic durations.

What we have not considered here is that some of these activities may be partially correlated; a schedule overrun in one activity may be indicative of a schedule overrun in another. We can use the same techniques as we did in Chapter 3.

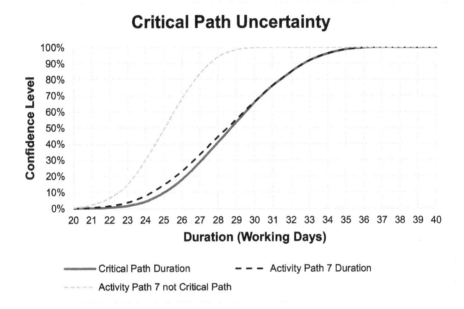

Figure 6.5 Critical Path Uncertainty

Critical Path Analysis and Monte Carlo Simulation are key components that allow us to perform a schedule assessment of Risk, Opportunity and Uncertainty, commonly referred to as Schedule Risk Analysis (SRA). To complete the SRA, we need to add any activity correlation and, of course any Risks or Opportunities that may (or may not) alter the activity durations or add/delete activities.

6.6   Chapter review

In this penultimate chapter, having recognised the need to plan or schedule, we dipped our feet into the heady world of Critical Path Analysis and we looked at how we can apply some simple logic with Binary Numbers to identify the Critical Path in a network of activities or tasks where we don't have access to specialist scheduling applications.

We explored the ideas of earliest and latest completion dates which give us an indication of optimistic and pessimistic views of the schedule, which in turn allowed us to mull over how we can use this to get an indication of the impact of schedule variation on cost using the Marching Army Technique from Chapter 4.

Finally, bringing it all back into the world of random numbers, we looked at how we can use Monte Carlo Simulation to give us a view of potential finish dates where the

Critical Path and the Critical Path duration are variable. Whilst our example was based on the plan for the estimating process, the principles extend to the project in question to be estimated, and as such forms an essential element of a Schedule Risk Analysis. As we discussed in Chapter 4, Schedule Risk Analysis can be a key component of a Top-down pessimistic perspective on the cost impact of Risk, Opportunity and Uncertainty Analysis.

References

Augustine, NR (1997) *Augustine's Laws (6th Edition)*, Reston, American Institute of Aeronautics and Astronautics, Inc.
Parkinson, CN (1955) 'Parkinson's Law', *The Economist,* November 19.

7 Finally, after a long wait ... Queueing Theory

This final chapter in the final volume provides an introduction to the delights of Queueing Theory. Sorry, if you've had to wait too long for this.

By the way 'Queueing' is not a spelling mistake or typographical error; in this particular topic area of mathematics and statistics, the 'e' before the '-ing' is traditionally left in place whereas the normal rules of British Spelling and Grammar would tell us to drop it ... unless we are talking about the other type of cue which would usually be 'cueing' but paradoxically could also be 'cuing' (Stevenson & Waite, 2011)!

Let's just pause for a moment and reflect on some idiosyncrasies of motoring. Did we ever:

- Sit in a traffic jam on the motorway, stopping and starting for a while, thinking, '*There must have been an accident ahead,*' but when you get to the front of the hold-up, there's nothing there! No angry or despondent drivers on the hard shoulder; not even a rubber-necking opportunity on the other side of the carriageway! It was just a 'Phantom Bottleneck' that existed just to irritate us.
- Come across those Variable Speed Limits (also on the motorways) and wonder, '*How do these improve traffic flow?*' How does making us go slower enable us to move faster?
- Notice how in heavy traffic the middle lane of a three-lane motorway always seems to move faster than the outside lane ... and that the inside lane sometimes moves faster still?
- Join the shortest queue at the toll booths, and then get frustrated when the two lanes on either side appear to catch up and overtake us ... until we switch lanes, of course, and then the one we were in starts speeding up?

All of these little nuances of motoring life are interlinked and can be explained by Queueing Theory, or at least it can pose a plausible rationale for what is happening.

However, Queueing Theory is not just about traffic jams and traffic management; it can be applied in other situations equally as well. Much of the early work on Queueing Theory emanates from Erlang (1909, 1917), who was working for the Copenhagen Telephone Company (or KTAS in Danish); he considered telephone call 'traffic' and how it could be better managed. Other examples to which Queueing Theory might be applied include:

- Hospital Accident & Emergency Waiting Times
- Telephone Call Centres
- Station Taxi Ranks
- Passport Applications
- Appointment Booking Systems
- Production Work-in-Progress
- Repair Centres

They all have some fundamental things in common. The demands for service, or arrivals into the system, are ostensibly random. There is also some limitation (capability and capacity) on how the system processes the demand and the number of 'channels' or requests for demand that that can be processed simultaneously.

Figure 7.1 depicts a simple queueing system but there could be different configurations in which we have multiple input channels, multiple queues and multiple fixed service points.

For example, let's consider a crowd queueing for entrance to a football match, or any other large spectator crown sporting event ...

- Multiple Input Channels: people may arrive individually on foot or by car, or in twos, threes or fours by the same way, or people may arrive in bigger batches by public or private transport such as buses, coaches or trains
- Multiple Queues and Service Points: there will be several access points or turnstiles around the stadium each with their own queue

It is costly having to queue but it is also costly to have excess capacity to service that queue (Wilkes, 1980, p.132). Queueing Theory allows us to seek out that optimum position ... the least of two evils! That said, the capacity of the system to process the demand

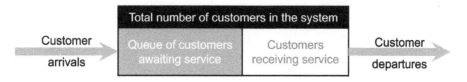

Figure 7.1 Simple Queueing System

must exceed the average rate at which the demand arises, otherwise once a queue has been formed, the system would not be able to clear it.

7.1 Types of queues and service discipline

We can divide the concept of queues into four distinct system groups:

- One-to-One System   Single ordered queue with one channel serving that queue (*Simple Queue*)
- One-to-Many System   Single ordered queue with more than one channel serving that queue
- Many-to-Many System Multiple queues with multiple channels. (*Perhaps this would be better dealt with by understanding of the mathematics of Chaos Theory? Don't worry, we are not going there.*) In reality we are often better dealing with these as multiple One-to-One Systems
- Many-to-One System   Multiple queues but only one channel to serve them all (*Survival of the Fittest!*). Think of a 3-lane motorway converging to a single lane due to an 'incident'

> ### A word (or two) from the wise?
>
> *'An Englishman, even if he is alone, forms an orderly queue of one.*
> **'George Mikes** (1946)
> Hungarian Author
> 1912–1987

Figures 7.2 to 7.5 show each of these configurations. In the latter two, where we have multiple queues, we might also observe 'queue or lane hopping'. Not shown in any of these is that each Queue might experience customer attrition where one or more customers decide that they do not want or have time to wait and are then lost to the system.

Recognising these complexities, Kendall (1953) proposed a coding system to categorise

Figure 7.2 Single Queue, Single Channel System

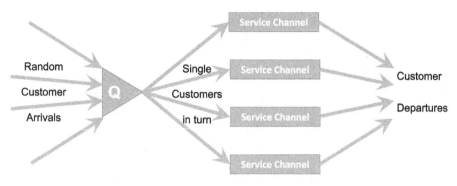

Figure 7.3 Single Queue, Multiple Channel System

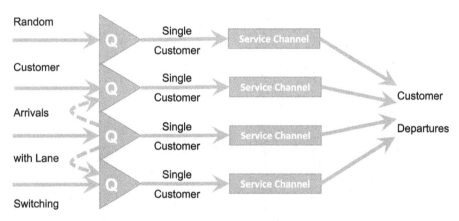

Figure 7.4 Multiple Queue, Multiple Channel System (with Queue Hoppers!)

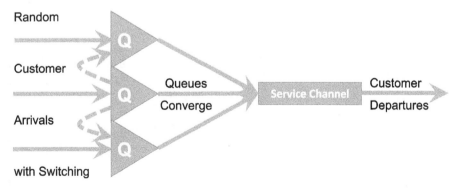

Figure 7.5 Multiple Queue, Single Channel System (with Queue Hoppers!)

Table 7.1 Kendall's Notation to Characterise Various Queueing Systems

A/S / c/K / N/D

Code	Meaning	Examples
A	Type of **Arrival** Process	Categories include (but not limited to):
		M for **Markovian** or **Memoryless** arrival process (see Section 7.2) or Poisson Distribution (Volume II Chapter 4 Section 4.9)
		E for an **Erlang** Distribution
		G for a **General** Distribution, which generally means another pre-defined Probability Distribution
		D for Degenerate or **Deterministic** Fixed Time interval
S	Type of **Service** Process	Similar to Arrival but in relation to the Service or Departure times
c	Number of Service **Channels**	The number of service channels through which the service is delivered, where c is a positive integer
K	System Capacity (**Kapacitat**)	The number of customers that the system can hold, either in a queue or being served. Arrivals in excess of this are turned away. If the parameter is excluded, the capacity is assumed to be unlimited
N	Size (**Number**) of Potential Customer Population	If omitted, the population is assumed to be unlimited. If the size of the potential customer population is small, this will have an impact on the Arrival Rate into the system (which will eventually be depleted)
D	Service **Discipline** applied to the queue	FIFO or First In First Out – Service is provided in the order that the demand arrived. *Sometimes expressed as First Come First Served (FCFS)*
		LIFO or Last In First Out – Service is provided in the reverse order to which the demand arose. *Sometimes expressed as Last Come First Served (LCFS)*
		SIPO or Served in Priority Order – Service is provided according to some pre-ordained priority
		SIRO or Served in Random Order – Service is provided in an ad hoc sequence

the characteristics of different queueing systems, which has now become the standard convention used. Although now extended to a six-character code, his original was based on three characters, written as A/S/c (*The last of the three being written in lower case!*). The extended 6-character code is A/S/c/K/N/D. Table 7.1 summarises what these are and gives some examples.

The relevance of the Server Discipline is one that relates to the likelihood of 'drop outs' from a system. If the customers are people, or perhaps to a lesser extent, entities owned by individuals, then they may choose to leave the system if they feel that they have to wait longer than is fair or reasonable. The same applies to inanimate objects which are 'lifed' such as 'Sell By Dates' on perishable products:

> FIFO is often the best discipline to adopt where people are involved as it is often seen as the fairest way of dealing with the demands on the system. The LIFO option can be used where the demand is for physical items that do not require a customer to be in attendance, such as examination papers in a pile awaiting to be marked. It doesn't matter if more are placed on top and served first so long as the last to be marked is achieved before the deadline.

SIPO is typically used in Hospital Accident & Emergency Departments. The Triage Nurse assesses each patient in terms of the urgency with which that patient needs to be seen in relation to all other patients already waiting. Another example would be Manufacturing Requirements Planning systems which assign work priorities based on order due dates. An example of SIRO is 'like parts' being placed in a storage bin and consumed as required by withdrawing any part by random selection from the bin for assembly or sale. This is the principle underpinning Kanban.

If the Discipline is not specified, the convention is to assume FIFO (*rather than SIRO which some might say is the natural state where there is a lack of discipline!*).

Finally, we leave this brief discussion of the types of queue discipline with the thoughts of George Mikes (1946), who alleged that an Englishman will always queue even when there is no need, but by default we must conclude that this is in the spirit of FIFO!

7.2 Memoryless queues

A fundamental characteristic of many queueing systems is that they are 'memoryless' in nature. The concept refers to the feature that the probability of waiting a set period of time is independent of how long we have been waiting already. For example, when we arrive in a queue, the probability of being served in the next ten minutes is the same as the probability of being served in the next ten minutes when we have already been waiting for 20 minutes! (*This might be taken as suggesting that we shouldn't bother waiting as we will probably never be served ... which always seems to be the feeling when we are in a hurry, doesn't it?*) However, what it really implies is that in Queueing Theory, previous history does not always help in predicting the future!

To understand what it means to be memoryless in a probabilistic sense, we first need to cover what Conditional Probability means. (*Don't worry, we're not doing a deep dive into this, just dipping a single toe in the murky waters of probability theory.*) For this we can simply turn to Andrey Kolmogorov (1933) and take his definition and the relationship he gave the statistical world:

For the Formula-philes: Conditional probability relationship (Kolmogorov)

Consider the probability of an event A occurring given that event B occurs also.

The conditional probability of A given B, $Pr(A \mid B)$, can be expressed as the probability of both A and B, $Pr(A\ B)$, occurring divided by the probability of B occurring independently, $Pr(B)$:

$$Pr(A \mid B) = \frac{Pr(A \cap B)}{Pr(B)}$$

For the Formula-philes: Definition of a memoryless Probability Distribution

A distribution is said to be memoryless if the probability of the time to the next event is independent of the time already waited. Consider the probability of an event occurring $F(t)$ within a given duration, T, independent of how long, S *(the Sunk Time)*, that we have waited already.

The Probability Distribution is said to be memoryless if:

$$F(t > S+T \mid t > S) = F(t > T) \qquad (1)$$

Substituting the Kolmogorov Conditional Probability Relationship in the left-hand side of (1) we get:

$$\frac{F([t > S+T] \cap [t > S])}{F(t > S)} = F(t > T) \qquad (2)$$

As $T > 0$, there are no circumstances where t can be greater than S+T and not greater than S, implying that:

$$F([t > S+T] \cap [t > S]) = F(t > S+T) \qquad (3)$$

Substituting (3) in (2) and re-arranging:

$$F(t > S+T) = F(t > S)F(t > T) \qquad (4)$$

... a memoryless probability distribution is one in which the probability of waiting longer than the sum of two values is the product of the probabilities of waiting longer than each value in turn

☼ For the Formula-phobes: Exponential growth or decay is memoryless?

Consider an exponential function in which there is an assumption of a constant rate of change. This could be radioactive decay, or a fixed rate of escalation over time. In terms of Radioactive substance, its Half-Life is the time it takes for its level of radioactivity to halve.

It doesn't matter where in time we are, the time it takes to halve its level is always the same ... it's almost as if it has forgotten (*or doesn't care*) where it is in time, or what level it was in the past when it started to decay, its only knows how long it will be until it halves again, which is easy because it is constant.

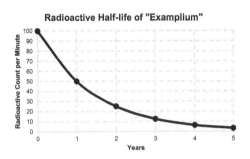

Radioactive Half-life of "Examplium"

Radioactive Decay, constant-rate Cost Escalation etc. are both examples of Exponential Functions. This Lack-of-Memory property (or memorylessness) also applies to the Exponential Probability Distribution.

This state of 'memorylessness' (*or lack-of-memory property if we find the term 'memoryless-ness' a bit of a tongue-twister*) can be applied to:

- The length of time a customer has to wait until being served
- The length of time between consecutive customer arrivals into a queue
- How long a server has to wait for the next customer (when there is no queue)

This state of memorylessness in queues leads us to the Exponential Distribution, the only continuous distribution with this characteristic (Liu, 2009) and the Geometric Distribution is the only Discrete Probability Distribution with this property.

If we refer back to Volume II Chapter 4 on Probability Distributions, we will notice the similarity in the form of the discrete Poisson's Probability Mass Function (PMF) and the continuous Exponential Distribution's Probability Density Function (PDF). Furthermore, if we assume that we have corresponding scale parameters for the two distributions, we will see that the Mean of one is the reciprocal of the other's Mean, indicating that the Mean Inter-Arrival Time can be found by dividing Time by the Mean Number of Customer Arrivals. Figure 7.6 illustrates the difference and the linkage between

For the Formula-philes: The Exponential Distribution is a memoryless distribution

Consider an Exponential Distribution with a parameter of λ.

The cumulative probability of the elapsed time t being less than or equal to T is:

$$F(t \leq T) = 1 - e^{-\lambda T} \qquad (1)$$

The cumulative probability of the elapsed time t being greater than T is the complement of (1):

$$F(t > T) = e^{-\lambda T} \qquad (2)$$

Similarly, the cumulative probability of being greater than S is:

$$F(t > T) = e^{-\lambda S} \qquad (3)$$

Similarly, the cumulative probability of being greater than $S+T$ is:

$$F(t > S + T) = e^{-\lambda(S+T)} \qquad (4)$$

Expanding (4):

$$F(t > S + T) = e^{-\lambda S} e^{-\lambda T} \qquad (5)$$

Substituting (2) and (3) in (5):

$$F(t > S + T) = F(t > S) F(t > T) \qquad (6)$$

… which is the key characteristic of a memoryless distribution previously demonstrated.

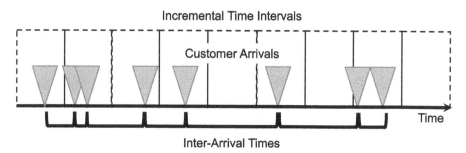

Incremental Time Intervals

Customer Arrivals

Time

Inter-Arrival Times

Figure 7.6 Difference Between Distribution of Arrivals and the Inter-Arrival Time Distribution

Customer Arrivals and their Inter-Arrival Times. (*Don't worry if Volume II Chapter 4 is not to hand, we've reproduced them here in the Formula-phile section.*

Mean Customer Arrival Rate = Number of Customers / Total Time
Mean Inter-Arrival Rate = Total Time / Number of Customers

We can go further than that and demonstrate that exponentially distributed inter-arrival times between successive Customer Arrivals gives us the equivalent of a Poisson Distribution for the number of arrivals in a given time period.

For the Formula-philes: Exponential Inter-Arrivals are equivalent to Poisson Arrivals

Consider a Poisson Distribution and an Exponential Distribution with a common scale parameter of λ,

Probability of C customer arrivals $p(C)$ in a time period, t, is given by the Poisson Distribution:

$$p(C) = \frac{\lambda^C}{C!} e^{-\lambda} \qquad (1)$$

The Mean of a Poisson Distribution is its parameter, λ:

$$\text{Average Customer Arrivals} = \lambda \qquad (2)$$

Probability of the Inter-Arrival Time $f(t)$ being t is given by the Exponential Distribution:

$$f(t) = \lambda e^{-\lambda t} \qquad (3)$$

The Mean of an Exponential Distribution is the reciprocal of its parameter

$$\text{Average InterArrival Time} = \frac{1}{\lambda} \qquad (4)$$

From (2) and (4) we can conclude that on average:

$$\text{InterArrival Time} = \frac{1}{\text{Customer Arrivals}} \qquad (5)$$

which shows that the Average Inter-Arrival Time in a Time Period is given by the reciprocal of the Average Number of Customer Arrivals in that Time Period.

We can do this very effectively using Monte Carlo Simulation. Figure 7.7 illustrates the output based on 10,000 iterations with an assumed Customer Arrival Rate of 1.6 customers per time period (or an average Inter-Arrival Rate of 0.625 time periods, i.e. based on the average Inter-Arrival Rate being the reciprocal of the average Customer Arrival Rate).

In this example, we have plotted the number of arrivals we might expect in period 4 based on the Monte Carlo model of random inter-arrivals with a theoretical Poisson Distribution. We would have got the similar results had we elected to show Period 7, or 12 etc.

7.3 Simple single channel queues (M/M/1 and M/G/1)

As we have not specified the K/N/D parameters, it is assumed that these are ∞, ∞, and FIFO. The main assumption here is that customers arrive at random points in time, and that the time spent in providing the service they want is also randomly distributed.

There are a number of well-trodden standard results that we can use from a simple queueing system. Let's begin by defining some key variables we may want from such a simple One-to-One Queueing System:

λ Average rate at which Customers arrive into the System at time, t

μ Average rate at which Customers are processed and exit the System at time, t

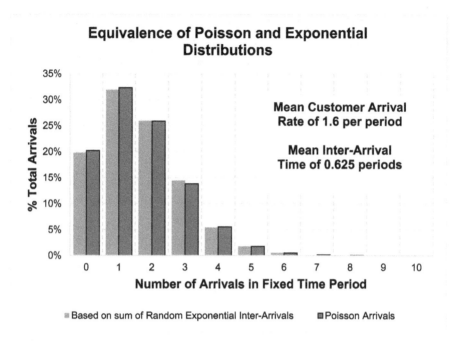

Figure 7.7 Equivalence of Poisson and Exponential Distributions

Q   Average Queue length, the number of Customers awaiting to be served at time, t
q   Average Waiting time that customers are held in in the Queue at time, t
R   Average Number of Customers in process of Receiving service at time, t
r   Average time that customers spend in Receiving service at time, t
S   Q + R, Average Number of Customers in the System at time, t
s   q + r, Average time that customers are in the System at time, t

Little (1961) made a big contribution to Queueing Theory with what has become known as **Little's Law**, (or sometimes known as Little's Lemma), that in a steady state situation:

Number of Customers in the System = Arrival Rate × Average Time in the System

This relationship holds for elements within the System, such as the queueing element. There is an assumption, however, that the average rate of providing the service is greater than the average Customer Arrival rate otherwise, the queues will grow indefinitely.

These relationships also hold regardless of the Arrival process... it doesn't have to be Poisson.

There is often an additional parameter defined that to some extent simplifies the key output results of Little's Law that we have just considered. It defines what is referred to as the Traffic Intensity (ρ) (sometimes referred to as the Utilisation Factor), and expresses the ratio of the Customer Arrival Rate (λ) to the Customer Service Rate (μ):

For the Formula-philes: Little's Law and its corollaries

For a Simple Queueing System, with a Mean Arrival rate of λ, and the Average Number of Customers being served is £ 1 (single service channel), Little's Law tells us that:

The total number of Customers, S, spending an average
time of s in the system: $S = \lambda s$ (1)

The total number of Customers, Q, waiting for service
with an average waiting time of q: $Q = \lambda q$ (2)

The total number of Customers, R, receiving service
with an average service time of r: $R = \lambda r$ (3)

By definition, the Average Time Receiving Service,
r, is the reciprocal of the Average Rate of Service completion. $r = \dfrac{1}{\mu}$ (4)

Substituting (4) in (3): $R = \dfrac{\lambda}{\mu}$ (5)

The average number of customers Receiving Service, R,
is the difference between the average of those in
the System, S and the average of those Queueing, Q: $R = S - Q$ (6)

The System Flow Rate is the difference between the
Service Rate and the Arrival Rate, but is also the
reciprocal of the average time spent in the system, s: $\mu - \lambda = \dfrac{1}{s}$ (7)

Substituting (7) in (1), the average number of customers in
the system: $S = \dfrac{\lambda}{\mu - \lambda}$ (8)

From (5) (6) and (8), the average Queue length is: $Q = \dfrac{\lambda}{\mu - \lambda} - \dfrac{\lambda}{\mu}$ (9)

Simplifying (9): $Q = \dfrac{\lambda^2}{\mu(\mu - \lambda)}$ (10)

Traffic Intensity = Customer Arrival Rate / Customer Service Rate

Note: If the system is not in a steady state, we can consider breaking it down into a sequence of steady state periods

For the Formula-philes: Equivalent expressions for single server systems

Various characteristics of the Queueing System can be expressed in terms of the arrival rate, service rate or the Traffic Intensity, ρ:

Traffic Intensity, or Utilisation Factor:

$$\rho = \frac{\lambda}{\mu} < 1$$

Average number of Customers receiving service, R:

$$R = \frac{\lambda}{\mu} \qquad R = \rho$$

Average Time that Customers spend in Receiving Service, r:

$$r = \frac{1}{\mu} \qquad r = \frac{\rho}{\lambda}$$

Average number of Customers in the System, S:

$$S = \frac{\lambda}{\mu - \lambda} \qquad S = \frac{\rho}{1 - \rho}$$

Average Time that Customers spend in the System, s:

$$s = \frac{1}{\mu - \lambda} \qquad s = \frac{1}{(1 - \rho)}$$

Average number of Customers in the Queue, Q:

$$Q = \frac{\lambda^2}{\mu(\mu - \lambda)} \qquad Q = \frac{\rho^2}{1 - \rho}$$

Average Time that Customers spend waiting in the Queue, q:

$$q = \frac{\lambda}{\mu(\mu - \lambda)} \qquad q = \frac{\rho}{(1 - \rho)}$$

These alternative relationships also highlight some other useful relationships we can keep in our back pockets (*we never know when they'll crop up at Quiz Night*):

For the Formula-philes: Other useful relationships for single server systems

Other useful relationships include:

Probability of there being at least S customers in the system, $F(S)$:

$$F(S) = \rho^S$$

Probability of there being S customers in the system,
$f(S)$ where $f(S) = F(S) - F(S-1)$:

$$f(S) = \rho^s (1-\rho)$$

Average Queue Length when there is a
queue (excluding zero length queues):

$$Q_{>0} = \frac{\mu}{\mu - \lambda}$$

Average Waiting Time as a function of Total System Time:

$$q = \rho s$$

Average number of customers in the System
as a function of Waiting Time:

$$S = \mu q$$

These relationships are all formulated on the premise that the average rate of service delivery (or capacity) is greater than the rate at which demand arises. If the service delivery or processing rate is temporarily less than the customer arrival rate, then a queue will be created. Let's look at such an example.

In the introduction to this section on Queueing Theory we mused about phantom bottlenecks on the motorway: where traffic is heavy and then stops, before progressing in a start-stop staccato fashion ... and then when we get to the front of the queue ... nothing. We expected to find a lane closure or signs of a minor collision, but nothing ... what was all that about?

For the Formula-phobes: Phantom bottlenecks on motorways

Some of us may be thinking '*How does all this explain phantom bottlenecks on the motorway?*' Before we explain it from the Queueing Theory perspective, let's consider the psychology of driving, especially in the outside or 'fast lane' as it is sometimes called.

In order to do that, let's just set the scene a little more clearly, and consider the psychology involved. Let us suppose that vehicles are being driven close together, in a somewhat 'tight formation' to prevent the lane switchers. Experiential observation will tell us that the gaps left between the vehicles is often much less than the recommended distance for safe reaction and braking time in the Highway Code. (*Note: This is not a Health and Safety example, but a reflection of a potential reality.*)

Everyone in one lane is driving at about the same speed until a lane-switcher forces their vehicle into the small gap between two other vehicles. (*Don't we just love them: big cars aimed at small spaces?*)

The driver of the vehicle to the rear instinctively brakes sharply, or at least feathers the brakes, causing the driver of the vehicle behind to brake also, probably

(Continued)

a little harder. If the gaps between the vehicles are small, this is likely to create a domino effect of successive drivers braking equal to or slightly more than the one in front. Eventually someone stops … as do all the vehicles that are behind if the gaps between them are also small … which they always seem to be in heavy traffic as everyone seems intent on mitigating against the lane-switchers by keeping the distance as small as they dare.

Note: This example does not advocate or condone driving closer to the vehicle in front than the recommended safe distance advised in the Highway Code (Department for Transport – Driving Standards Agency, 2015).

Let's look at this from a Queueing Theory perspective. We are going to work in miles per hour.

Rule of Thumb

The nice thing about working in miles per hour is that it gives us a simple rule of thumb for yards per second. To convert from miles to yards requires us to multiply by 1,760. To convert 'per hour' to 'per second' we must divide by 3,600. This gives us a net conversion factor of 0.489 … which as a rule of thumb can be interpreted as almost a half. So an average speed of 50 mph is approximately 25 yards per second.

[If we assume that on a congested motorway where the average speed is around 50 mph, and that cars are some 20 yards apart (*yes, we could argue that this is conservatively large when we think how close some people drive in such circumstances*), then coupled with the assumption also that the average car length in Europe is around 4.67 yards, then this suggests that each vehicle occupies around 25 yards of 'personal space' … or equivalent to one second reaction time.

If a car slows down, this one second reaction time is equivalent to a shorter distance, and induces perhaps a slower speed in the driver behind of around 10 mph.

We get the result we can see in Figure 7.8, which shows a snapshot of the cars passing through the 25-yard bottleneck where our example lane switcher committed the act. (*Come on admit it, wouldn't we all love to make an example of them?*)

The queue will only dissipate if either the rate at which cars can be processed through the section increases (because the road in front of them is clear), or if the rate at which vehicles arrive at the bottleneck reduces.

7.3.1 Example of Queueing Theory in action M/M/1 or M/G/1

Using the standard steady state equations can be less than straightforward, so it is often easier to model the results using a simulation tool. Here we will use a Monte Carlo approach in Microsoft Excel.

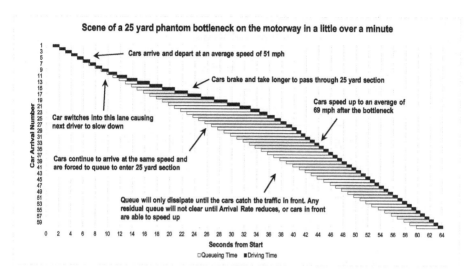

Figure 7.8 The Curse and Cause of Phantom Bottlenecks on Motorways

In this example, we will consider a very basic repair facility that can process one repair at a time (*and we can't get more basic than that*) and that it exists in a steady state environment (i.e. not growing or reducing).

Assuming Steady State Random Arrivals, the procedure we will follow will be:

1. Analyse the historical average rate at which the random arrivals occur.
2. Analyse the historical data for repair times and establish whether there is a predictable pattern or distribution of repair times.
3. Create a model to simulate a range of potential scenarios in order to create a design solution.

For simplicity, at the moment we will assume that steps 1 and 2 have been completed and that the arrivals are random but varying around a steady state average rate. Let us also assume initially that the repair times are exponentially distributed. This gives us a M/M/1 Queueing Model, illustrated in Table 7.2.

1. Cells C2 and H2 are the Average Repair Inter-Arrival Time and Repair Time respectively
2. Cells C3 and H3 are the Arrival and Repair Rates expressed as a number per month and are the reciprocals of Cells C2 and H2 respectively
3. Cell J3 is the theoretical Traffic Intensity or Utilisation expressed as the ratio of the Average Arrival Rate (Cell C3) and the Average Repair Rate (H3)
4. Cell K3 calculates the theoretical Average Queueing Time as Traffic Intensity or Utilisation (J3) divided by the difference in the Arrival Rate and Repair Rate

Table 7.2 Example Simple Repair Facility with a Single Service Channel and Unbounded Repair Times

	A	B	C	D	E	F	G	H	I	J	K
1										Theoretical Utilisation	Mean Queue Time (Mths)
2	Mean Inter-Arrival Time	0.50	Months			Mean Repair Time >	0.40	Months		80.0%	1.60
3	Mean Arrival Rate	2	per Month			Mean Repair Rate	2.50	per Month			
5		Repair Number	Months since last Repair Receipt	Arrival Time (Cum Months from Start)	Repair Receipt Month	Repair Start Time (Cum Months from Start)	Repair Start Month	Repair Time in Months	Completion Time (Cum Months from Start)	Repair Return Month	Queueing Time (Months)
6	Start >	0	0	0	0	0	0	0	0	0	0
7		1	0.443	0.443	1	0.443	1	0.367	0.830	1	0.000
8		2	1.017	1.460	2	1.460	2	0.143	1.603	2	0.000
9		3	0.252	1.712	2	1.712	2	0.389	2.101	3	0.000
10		4	0.042	1.754	2	2.101	3	0.083	2.183	3	0.347
11		5	0.147	1.901	2	2.183	3	0.377	2.560	3	0.282
12		6	0.094	1.995	2	2.560	3	0.460	3.020	4	0.565
13		7	0.247	2.242	3	3.020	4	0.103	3.123	4	0.778
14		8	0.224	2.466	3	3.123	4	0.666	3.789	4	0.657
15		9	0.164	2.630	3	3.789	4	1.042	4.832	5	1.160
16		10	0.213	2.843	3	4.832	5	1.146	5.977	6	1.989
17		11	0.021	2.864	3	5.977	6	0.609	6.587	7	3.114
18		12	0.323	3.186	4	6.587	7	0.025	6.612	7	3.400

(H3– C3). Using Little's Law we can easily determine the Average Queue Length by multiplying the Average Queueing Time by the Arrival Rate in Cell C3

5. Column C represents the random Poisson Arrivals for the Repairs which, as we discussed earlier, is equivalent to Exponentially distributed Inter-Arrival Times between the consecutive Repairs

Unfortunately, in Microsoft Excel there is no obvious function that returns the value from an Exponential Distribution for a given cumulative probability, i.e. there is no EXPON.INV function. However, from Volume II Chapter 4 we know that the Exponential Distribution is a special case of the Gamma Distribution, with an alpha parameter of 1.

The good news is that there is the equivalent function in Microsoft Excel to generate the Gamma Distribution values for a given cumulative probability. For example, the values in Column C can be generated by the expression **GAMMA.INV(RAND(), 1, $C$2)**. Exploiting this special case gives us the capability to generate the value for an Exponential Distribution

6. Column D is the running total of Column C giving us the simulated point in time that the repair is received

7. Column E is a 'nice to have' for reporting purposes and expresses the Month in which the repair arrives. We will look back at this shortly in Section 7.7 on the Poisson Distribution

8. Column F calculate the earliest Repair Start Time by calculating the Maximum of either the current Repair's arrival time from Column D or the previous Repair's Completion Time from Column I (which we will explain in Step 11)

9. Column G is similar to Column E but relates to the month in which the repair process is deemed to have started

10. Column H is assumed to be a random repair duration with an Exponential Distribution and is calculated as **GAMMA.INV(RAND(), 1, $H$2)** (see Step 5 for an explanation)
11. Column I is the time at which the repair is completed as is the sum of the Start Time (Column F) and the Repair Duration (Column H). The repair is ready to leave the system
12. Column J is similar to Column E but relates to the month in which the Repair is deemed to have been completed
13. Column K calculates the Queueing Time for each Repair as the difference between the Repair Start Time (Column F) and its Arrival Time into the system (Column D)

Figure 7.9 illustrates a potential outcome of the model, highlighting that queues tend to grow when there is an excessive random repair time and diminishes when the repair time is around or less than the average, or when the Arrival Rate reduces to less than its average. Figure 7.10 gives us a view on the observed Queue Lengths in this model in comparison with the Average Queue Length derived using Little's Law.

Let's look at our model again with the same Arrival pattern but with a different assumption on the distribution of repair times. What if our analysis showed that the repair time was a Beta Distribution with parameters $\alpha=2$ and $\beta=11.33$, with a minimum repair time of one week (0.25 months) and a maximum of 1.25 months? This would give

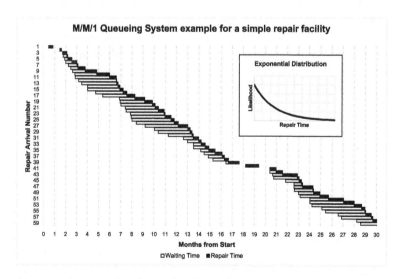

Figure 7.9 Example of a Single Channel Repair Facility with an Unbounded Repair Time Distribution

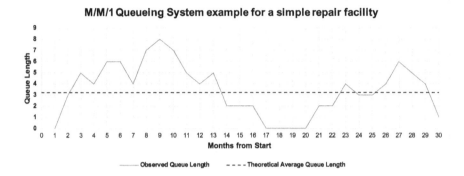

Figure 7.10 Example Queue Lengths for a Single Channel Repair Facility with Unbounded Repair Times

us the same Average Repair Time of 0.4 months that we used in the previous example. Using Kendall's classification, this is now a M/G/1 Queueing System because we have now replaced the memoryless (M) Exponential Distribution used for the service function with a known general (G) distribution, in this case a Beta Distribution. Table 7.3 illustrates the model set-up. The only difference to the previous M/M/1 model is in the calculation of the random repair time which now utilises the Microsoft function for the Beta Distribution: **BETA.INV(RAND(), *alpha, beta, minimum, maximum*)**.

To enable a direct comparison, we have recycled the same set of random numbers to generate both the Arrival Times and the Repair Times; the only difference therefore is the selected value of the Repair Time. *(Such recycling smacks of the modelling equivalent of being an eco-friendly Green!)*

Figures 7.11 and 7.12 provide the comparable illustrations of this model with those presented for the M/M/1 model. Whilst we still get queues building up, there is less chance of extreme values because we have limited the maximum Repair Time to nine weeks (2.25 months).

The problem with a Beta Distribution (as we discussed in Volume II Chapter 4, and is evident from the insert in Figure 7.11) is that the long tail means that whilst larger values from the Beta Distribution are theoretically possible, they are extremely unlikely. In essence, in this example the realistic range of Repair Times is between 0.25 and 1.0 months. In 10,000 iterations of this model, the maximum Repair Time selected at random was around 1.0. We should bear this in mind that if we our basing our Repair Time Distribution on empirical data we should be wary of assigning long tail Beta Distributions with high value parameters. In this case, the Beta parameter of 24.67 is probably excessively large. (In this example, the values were chosen to maintain the Average Repair Time within a perceived lower and upper bound. Normally, the parameter values

Table 7.3 Example of Simple Repair Facility with a Bounded Repair Time Distribution

	A	B	C	D	E	F	G	H	I	J	K
1							alpha	2			
2							beta	24.67			
3						Min Repair Time >		0.25	Months		
4						Max Repair Time >		2.25	Months	Theoretical	Mean Queue
5	Mean Inter-Arrival Time	0.50	Months			Mean Repair Time >		0.4	Months	Utilisation	Time (Mths)
6	Mean Arrival Rate	2	per Month			Mean Repair Rate		2.50	per Month	80.0%	1.60
7											
8		Repair Number	Months since last Repair Receipt	Arrival Time (Cum Months from Start)	Repair Receipt Month	Repair Start Time (Cum Months from Start)	Repair Start Month	Repair Time in Months	Completion Time (Cum Months from Start)	Repair Return Month	Queueing Time (Months)
9	Start >	0	0	0	0	0	0	0	0	0	0
10		1	0.443	0.443	1	0.443	1	0.410	0.853	1	0.000
11		2	1.017	1.460	2	1.460	2	0.335	1.795	2	0.000
12		3	0.252	1.712	2	1.795	2	0.411	2.206	3	0.084
13		4	0.042	1.754	2	2.206	3	0.312	2.518	3	0.452
14		5	0.147	1.901	2	2.518	3	0.407	2.925	3	0.617
15		6	0.094	1.995	2	2.925	3	0.429	3.354	4	0.930
16		7	0.247	2.242	3	3.354	4	0.320	3.674	4	1.112
17		8	0.224	2.466	3	3.674	4	0.480	4.154	5	1.208
18		9	0.164	2.630	3	4.154	5	0.562	4.716	5	1.524
19		10	0.213	2.843	3	4.716	5	0.583	5.299	6	1.873
20		11	0.021	2.864	3	5.299	6	0.466	5.765	6	2.436
21		12	0.323	3.186	4	5.765	6	0.281	6.047	7	2.579

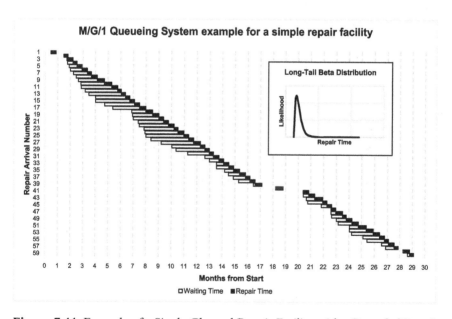

M/G/1 Queueing System example for a simple repair facility

Long-Tail Beta Distribution

Likelihood

Repair Time

Repair Arrival Number

Months from Start

□ Waiting Time   ■ Repair Time

Figure 7.11 Example of a Single Channel Repair Facility with a Bounded Repair Time Distribution

would be calibrated around a thorough analysis of actual repair times, and not just three statistics of Minimum, Mean and Maximum.) When modelling with distributions such as the Beta Distribution, it is advisable to plot the Cumulative Distribution Function (CDF) that we intend to use just to check that it gives us the realistic range that we expect.

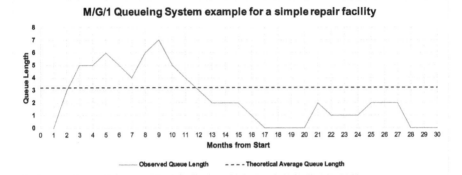

Figure 7.12 Example Queue Lengths for a Single Channel Repair Facility with Bounded Repair Times

7.4 Multiple channel queues (M/M/c)

This is where life begins to get a little rough, in terms of the number juggling required, (*although in practice there may also be a bit of argy-bargy and jostling for position in the queues*).

These systems are appropriate where the average service time is greater than the average Customer Inter-Arrival Time. By having a number of service channels we can dilute the effective average Arrival Rate to each service channel to being the true Arrival Rate divided by the number of service channels. If we have four channels then we would expect that each channel would serve approximately every fourth customer. If we have multiple physical queues rather than a single queue (physical or virtual) then this can lead to queue discipline issues with queue hopping as we illustrated previously in Figures 7.4 and 7.5. A virtual queue might be one where customers take a counter ticket which gives them their service sequence number. (*We may have noticed that we often appear to choose the wrong queue ... how often have we elected to switch queues or seen others do it ... and then find it makes little difference?*)

Little's Law still applies to multiple channel systems, although we do need to amend some of the standard relationships when we are in a steady state situation. As this is meant to be just an introduction to Queueing Theory, we will not attempt to justify these formulae here. They can be found in other reputable books on the subject and in Operational Research books (e.g. Wilkes, 1980).

... like I said, things would begin to get a little rough on the formula front with Multi-Channel Queues.

For the Formula-philes: Useful Relationships for Multi-Server Systems

Some potentially useful relationships with Multi-Channel Queueing Systems in which Little's Law still applies, include:

Traffic Intensity:
$$\rho = \frac{\lambda}{c\mu} \quad (1)$$

Probability P_0 of there being no customers in the system
$$P_0 = \frac{c!(1-\rho)}{(\rho c)^c + c!(1-\rho)\left(\sum_{n=0}^{c-1}\frac{1}{n!}(\rho c)^n\right)} \quad (2)$$

Probability of there being c or more customers in the system (i.e. the probability of having to wait for service)
$$P_c = \frac{(c)^c}{c!(1-\lambda)}P_0 \quad (3)$$

Average Number of customers, S in the system:
$$S = \frac{\rho(\rho c)^c}{c!(1-\rho)^2}P_0 + \rho c \quad (4)$$

Average Number of customers, Q in the Queue:
$$Q = \frac{\rho(\rho c)^c}{c!(1-\rho)^2}P_0 \quad (5)$$

If s is the average time a customer spends in the system, then from Little's Law:
$$S = \lambda s \quad (6)$$

Substituting (1) and (4) in (6) and re-arranging:
$$s = \frac{(\rho c)^c}{c!(1-\rho)^2 c\mu}P_0 + \frac{1}{\mu} \quad (7)$$

If q is the average time a customer spends waiting for service, then from Little's Law:
$$Q = \lambda q \quad (8)$$

Substituting (1) and (5) in (8) and re-arranging:
$$q = \frac{(\rho c)^c}{c!(1-\rho)^2 c\mu}P_0 \quad (9)$$

Rather than try to exploit these formulae, we shall use the basic principles of modelling the scenarios in Microsoft Excel.

7.4.1 Example of Queueing Theory in action M/M/c or M/G/c

Assuming steady state Random Arrivals, the procedure we will follow will be:

1. Analyse the historical average rate at which the random arrivals occur.
2. Analyse the historical data for Repair Times and establish whether there is a predictable pattern or distribution of Repair Times.

3. Create a model to simulate a range of potential scenarios in order to create a design solution.

For simplicity, at the moment we will assume that steps 1 and 2 have been completed and that the arrivals are random but vary around a steady state average. Let us also assume initially that the repair times are exponentially distributed (memoryless) but on average their Arrival Rate is greater than the expected Average Repair Times. This means that we will need more repair service channels to satisfy the demand. Table 7.4 illustrates a potential M/M/c Queueing Model. Let's begin with the same repair arrival rate that we assumed in our earlier models in Section 7.3, where we assumed an Average Arrival Rate of two repairs per month. This time we will assume a Mean Repair Time of 1.2 months, equivalent to a capacity of 2.5 repairs per month across three service channels or stations (calculated by taking the Number of Repair Stations divided by Average Repair Time.) Let's begin by assuming unbounded Exponential Repair Times.

The set-up procedure for this model is as follows:

1. Cells C2 and I2 are the Average Repair Inter-Arrival Time and Repair Time respectively
2. Cell F3 is a variable that will allow us to increase or decrease the number of service channels. In this particular model we have assumed three service channels or stations initially. (*There's a hint here that we are going to change it later*)
3. Cells C3 and I3 are the Arrival and Repair Rates expressed as a quantity per month. Cell C3 is the reciprocal of Cell C2. Cell I3 is calculated by dividing the number of Repair Stations, F3 by I2

Table 7.4 Example of a 3-Channel Repair Facility with an Unbounded Repair Time Distribution

	A	B	C	D	E	F	G	H	I	J	K	L
2	Mean Inter-Arrival Time		0.50	Months		No of Repair Stations	Mean Repair Time >		1.20	Months	Theoretical Utilisation	Mean Queue Time (Mths)
3	Mean Arrival Rate		2	per Month		3	Mean Repair Rate		2.50	per Month	80.0%	0.53
5		Repair Number	Months since last Repair Receipt	Arrival Time (Cum Months from Start)	Repair Receipt Month	Time of Next Available Repair Station	Repair Start Time (Cum Months from Start)	Repair Start Month	Repair Time in Months	Completion Time (Cum Months from Start)	Repair Return Month	Queueing Time (Months)
6	Start >	0	0	0	0	0	0	0	0	0	0	0
7		1	0.443	0.443	1	0.000	0.443	1	1.162	1.605	2	0.000
8		2	1.017	1.460	2	0.000	1.460	2	0.429	1.889	2	0.000
9		3	0.252	1.712	2	0.000	1.712	2	1.167	2.878	3	0.000
10		4	0.042	1.754	2	1.605	1.754	2	0.248	2.002	3	0.000
11		5	0.147	1.901	2	1.889	1.901	2	1.130	3.031	4	0.000
12		6	0.094	1.995	2	2.002	2.002	3	1.381	3.383	4	0.007
13		7	0.247	2.242	3	2.878	2.878	3	0.308	3.186	4	0.636
14		8	0.224	2.466	3	3.031	3.031	4	1.999	5.030	6	0.565
15		9	0.164	2.630	3	3.186	3.186	4	3.127	6.313	7	0.557
16		10	0.213	2.843	3	3.383	3.383	4	3.438	6.821	7	0.540
17		11	0.021	2.864	3	5.030	5.030	6	1.828	6.858	7	2.167
18		12	0.323	3.186	4	6.313	6.313	7	0.076	6.389	7	3.126

4. Column C represents the random Poisson Arrivals for the Repairs which, as we discussed earlier, is equivalent to being Exponentially distributed Inter-Arrival Times between the consecutive Repairs. (To save us looking back to Section 7.3, just in case you haven't committed it to memory, or jumped straight in here, the calculation rationale in Microsoft Excel is repeated below – *see how considerate I am to you!*)

In Microsoft Excel, there is no obvious function that returns the value from an Exponential Distribution for a given cumulative probability. However, from Volume II Chapter 4 we know that the Exponential Distribution is a special case of the Gamma Distribution, with its alpha parameter set to 1. We can exploit that here to generate the Exponential Distribution values for a given cumulative probability using the inverse of the Gamma Distribution. For example, the values in Column C can be generated by the expression **GAMMA.INV(RAND(), 1, $C$2)**

5. Column D is the running total of Column C to give the point in time that the repair is received
6. Column E is a 'nice to have' for reporting purposes and expresses the Month in which the repair arrives
7. Column F calculates the time of the next available Repair station based on the projected completion times of the previous repairs in Column J (which we will discuss in Step 12). We can do this using the special Microsoft Excel function **LARGE(*array, k*)** where **array** is the range of all repair completion times in Column J across all previous rows, and **k** is the rank order parameter, equal here to the number of Repair Stations. For example: **F12=LARGE($J$6:$J11,F$3)**
8. However, using this formula will create an error for the first few repairs (three in this example), until each repair station has processed at least one repair. To avert this we need to add a condition statement that sets the Time of the Next Available Repair Station (Column F) to zero if the Repair Number (Column B) is less than or equal to the number of Repairs Stations (Cell F3).

For example: **F12=IFERROR(LARGE($J$6:$J11,F$3),0)**

9. We use Column G to calculate the earliest Repair Start Time by calculating the Maximum of either the current Repair's arrival time from Column D if there is no queue, or the Time of the Next Available Repair Station from Column F
10. Column H is similar to Column E but relates to the month in which the repair process is deemed to have started
11. Column I is assumed to be a random repair duration with an Exponential Distribution and is calculated as **GAMMA.INV(RAND(), 1, $H$2)** (see Step 4 for an explanation)

12. Column J is the time at which the repair is completed as is the sum of the Start Time (Column G) and the Repair Duration (Column I)
13. Column K is similar to Column E but relates to the month in which the Repair is deemed to have been completed
14. Column L calculates the Queueing Time for each Repair as the difference between the its Start Time (Column F) and its Arrival Time (Column D)

Figure 7.13 illustrates a potential outcome of the model. From this we can see that queues begin to build when the Repair Times are elongated, but between the three stations, the queues are manageable with 80% Utilisation or Traffic Intensity. If we were to have the same demand but invested in a fourth Repair Station, we can reduce the queues significantly (but not totally) with an overall utilisation of 60% as illustrated in Figure 7.14.

We can repeat this multi-channel model as we did for the single channel scenario and replace the unbounded Exponential Repair Time Distribution with a bounded Beta Distribution. For a like-for-like comparison we will assume the same demand Arrival Rate and an Average Repair Time of some three times the single channel case, as shown in Table 7.5. The mechanics of the model set-up are identical to the M/M/3 example, with the exception of the calculation of the Repair Time in Column I which is now assumed to be a Beta Distribution with the parameters shown in Cells I1 to I4. A typical output is shown in Figure 7.15.

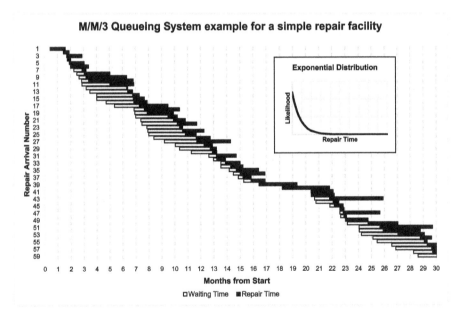

Figure 7.13 Example of a 3-Channel Repair Facility with an Unbounded Repair Time Distribution

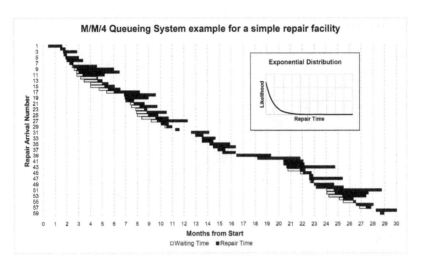

Figure 7.14 Example of a 4-Channel Repair Facility with an Unbounded Repair Time Distribution

Table 7.5 Example of a 3-Channel Repair Facility with a Bounded Repair Time Distribution

	A	B	C	D	E	F	G	H	I	J	K	L
1								alpha	2			
2								beta	4.00			
3							Min Repair Time >	0.3	Months		
4						No of Repair	Max Repair Time >	3	Months		Theoretical	Mean Queue
5	Mean Inter-Arrival Time	0.50	Months			Stations	Mean Repair Time >	1.2	Months		Utilisation	Time (Mths)
6	Mean Arrival Rate	2	per Month			3	Mean Repair Rate >	2.50	per Month		80.0%	0.53
7												
8		Repair Number	Months since last Repair Receipt	Arrival Time (Cum Months from Start)	Repair Receipt Month	Time of Next Available Repair Station	Repair Start Time (Cum Months from Start)	Repair Start Month	Repair Time in Months	Completion Time (Cum Months from Start)	Repair Return Month	Queueing Time (Months)
9	Start >	0	0	0	0	0	0	0	0	0	0	0
10		1	0.443	0.443	1	0.000	0.443	1	1.315	1.758	2	0.000
11		2	1.017	1.460	2	0.000	1.460	2	0.889	2.349	3	0.000
12		3	0.252	1.712	2	0.000	1.712	2	1.317	3.029	4	0.000
13		4	0.042	1.754	2	1.758	1.758	2	0.736	2.494	3	0.004
14		5	0.147	1.901	2	2.349	2.349	3	1.300	3.649	4	0.448
15		6	0.094	1.995	2	2.494	2.494	3	1.413	3.907	4	0.499
16		7	0.247	2.242	3	3.029	3.029	4	0.791	3.820	4	0.787
17		8	0.224	2.466	3	3.649	3.649	4	1.647	5.296	6	1.184
18		9	0.164	2.630	3	3.820	3.820	4	1.969	5.789	6	1.191
19		10	0.213	2.843	3	3.907	3.907	4	2.041	5.948	6	1.064
20		11	0.021	2.864	3	5.296	5.296	6	1.587	6.883	7	2.433
21		12	0.323	3.186	4	5.789	5.789	6	0.531	6.320	7	2.603

7.5 How do we spot a Poisson Process?

The answer to that one is perhaps '*with difficulty*'!

We have concentrated so far on the pure Poisson Arrival Process, but how would we know if that was a valid assumption with our data?

If the Average Repair Time is significantly less than the average Inter-Arrival Time, who cares? We'd just do the repair when it turned up, wouldn't we, because it would be so unlikely for there to be any queue to manage let alone theorise about?

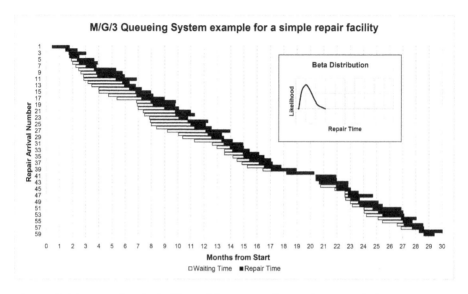

Figure 7.15 Example of a 3-Channel Repair Facility with a Bounded Repair Time Distribution

Let's consider the key characteristics of a Poisson Process:

- Arrivals are random
- Inter-Arrival Times are Exponentially Distributed
  - This implies that relatively smaller gaps between Arrivals are more likely to occur than larger gaps
  - The Standard Deviation of the Inter-Arrival Times are equal to the Mean value of those same Inter-Arrival Times
- There is no sustained increase or decrease in Arrivals over time
  - The best fit line through the number of Arrivals per time period is a flat line
  - The cumulative number of Arrivals over time is linear
- The incidence of Arrivals in any time period follow a Poisson Distribution
  - The Variance of the number of Arrivals in a time period is equal to the Mean of those number of Arrivals per time period

However, as with all data there will be variations in what we observe and there will be periods when there may seem to be contradictions to the above. However, we should resist the urge to jump to premature conclusions and have to fall back on a dose of statistical tolerance. Two things would help us out here:

1. Date and/or Time Stamping of the incoming data
2. A longer period of time over which we gather our empirical data

If all our demand arisings were date and time stamped on receipt, then we could calculate the Inter-Arrival Times and verify whether they were Exponentially Distributed or not, using the techniques we discussed in Volume III Chapter 7. (*Don't worry about looking back just yet, we will provide a reminder shortly.*) However, this very much depends on the specification of our in-house or commercial off-the-shelf information systems. In many instances, we may only be able to say which time period (week or month) the demand arose, which scuppers the Inter-Arrival Time approach to some degree if it is a relatively high rate per time period.

Let's look at an example that is in fact a Poisson Process but where we might be misled.

The demand input data that we used in our previous models in Sections 7.3.1 and 7.4.1 was generated specifically to be Poisson Arrivals. Suppose then instead of this being randomly simulated generated data, that it was genuine random data, but that it was not date/time stamped; instead we only know which month the demand arose.

Now let's look at the data we have after 12 months, 24 months and 48 months; our view of the underlying trend may change ... the trouble is we cannot usually wait four years before we make a business decision such as sizing our repair facility, or contractually committing to a maximum turn-around time. In Figure 7.16, we might conclude that there appears to be a decreasing linear trend over time. Whilst the Coefficient of Determination (R^2) is very low for the unit data (arisings per month), the cumulative perspective is a better fit to a quadratic through the origin, indicating perhaps a linear trend (see Volume III Chapter 2).

If we were to re-visit our data after two years (Figure 7.17), we would see that the overall rate of decrease was less than had been indicated a year earlier. In fact, it seems to have dipped and then increased. Perhaps the first year decreased as previously observed and then the second year flat-lined ... *or maybe there's some other straw at which we can clutch!*

If we'd had the luxury of four years' data (Figure 7.18), we would have found that the demand for repairs was reasonably constant, albeit with a random scatter around an average rate, and a straight line cumulative.

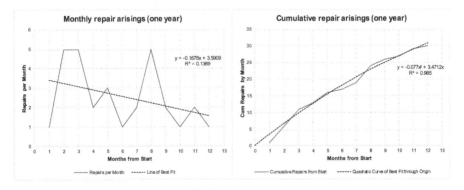

Figure 7.16 Apparent Reducing Demand for Rpairs Over 12 Months

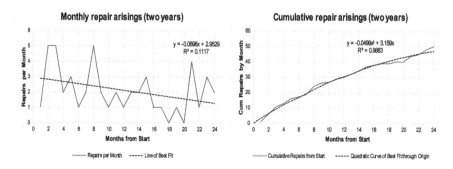

Figure 7.17 Apparent Reducing Demand for Repairs Over 24 Months

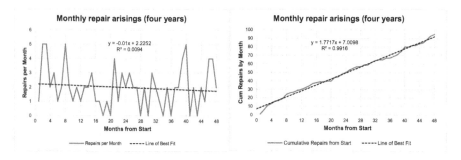

Figure 7.18 Constant Demand for Repairs Over 48 Months

What else could we have tried in order to test whether this was a Poisson Process at work? The giveaway clue is in the name, Poisson; the frequency of repair arisings in any month should be consistent with a Poisson Distribution. In Table 7.6 we look at the first 12 months:

1. The Observed % Distribution is simply the proportion of Observed Occurrences per Month relative to the total for 12 months

2. The Poisson % Distribution can be calculated in Microsoft Excel as **POISSON. DIST**(*RepairsPerMonth, MeanRepairs, FALSE*). (The FALSE parameter just specifies that we want the Probability Mass Function (PMF) rather than the Cumulative Distribution Function (CDF).) Here we will assume the observed sample's Mean Repairs per Month which can be calculated as the **SUMPRODUCT**(*RepairsPer-Month, ObservedMonthlyOccurrences*)**/Time Period.** (The sum of the first two columns multiplied together divided by 12)

Table 7.6 Comparing the Observed Repair Arisings per Month with a Poisson Distribution

Covering	12	months

Repairs per Month	Number of Observed Occurrences	Observed % Distribution	Poisson % Distribution	Number of Poisson Occurrences (Unrounded)	Cumulative % Observed Arisings	Cumulative % Poisson Arisings
0	0	0.0%	8.2%	0.985	0.0%	8.2%
1	4	33.3%	20.5%	2.463	33.3%	28.7%
2	4	33.3%	25.7%	3.078	66.7%	54.4%
3	1	8.3%	21.4%	2.565	75.0%	75.8%
4	0	0.0%	13.4%	1.603	75.0%	89.1%
5	3	25.0%	6.7%	0.802	100.0%	95.8%
6	0	0.0%	2.8%	0.334	100.0%	98.6%
7	0	0.0%	1.0%	0.119	100.0%	99.6%
8	0	0.0%	0.3%	0.037	100.0%	99.9%
9	0	0.0%	0.1%	0.010	100.0%	100.0%
10	0	0.0%	0.0%	0.003	100.0%	100.0%

Total	30		↑	↑		↑
Mean	2.50	>>> Poisson parameter		2.50		
Variance	2.64			1.581		
Std Dev	1.624					

3. The unrounded Number of Poisson Occurrences is simply the Time Period multiplied by the Poisson % Distribution to give us the number of times we can expect this Number of Repairs per Month in the specified time period

From our table, we may note that:

*   We never had a month without a repair arising (*zero occurred zero times*)
*   One and two repairs per month occurred four times each, both being greater than we would expect with a Poisson Distribution (*but hey, chill, we can expect some variation*)
*   Three repairs occurred just once
*   Perhaps surprisingly we had three months where five repairs turned up, and yet this was the maximum number also, whereas the Poisson Distribution was predicting one at best after rounding
*   The observed sample Mean and Variance are similar values, which would be consistent with a Poisson Distribution (see Volume II Chapter 4)
*   Despite these variations in monthly observations, from the Cumulative Distribution perspective in the right hand graph of Figure 7.19, it looks to be a not unreasonable fit to the data. (*Note the characteristic double negative of the statistician in me rather than a single positive*)

Rather than merely accept its visual appearance, in order to test the goodness of fit, we can perform a Chi-Squared Test (Volume II Chapter 6) on the Observed data against the

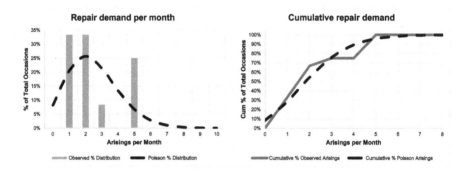

Figure 7.19 Comparing the Observed Repair Arisings per Month with a Poisson Distribution

Expected values derived from a Poisson Distribution based on the sample Mean. Unfortunately, even though there is a Microsoft Excel Function for the Chi Squared Test, we cannot use it here as this assumes that the number of degrees of freedom is one less than the number of observations; that doesn't work when we are trying to fit a distribution with unknown parameters.

In this case, with a Poisson Distribution, we have one unknown parameter, the Mean. We have only assumed it to be equal to the sample Mean; in reality it is unknown! *(If we had been testing for normality we would have had to deduct two parameters, one for each of the Mean and Standard Deviation.)*

Caveat augur

Degrees of freedom can be tricky little blighters. It is an easy mistake to make to use the Chi-Squared Test blindly 'out of the box', which assumes that the degrees of freedom are one less than the number of data points. This is fine if we are assuming a known distribution.

If we have estimated the parameters, such as the sample Mean of the observed data, then we lose one degree of freedom for each parameter of the theoretical distribution we are trying to fit.

Even though we can't use the **CHISQ.TEST** function directly in Excel, we can perform the Chi-Squared Test manually. (*Was that an audible groan, I heard? Come on, you know you'll feel better for it afterwards!*) as depicted in Table 7.7.

Table 7.7 Calculating the Chi-Squared Test for Goodness of Fit (Long Hand) over a 12 Month Period

Repairs per Month	Observed Occurrences (O)	Expected Poisson Occurrences (E)	Fit Statistic $\frac{(O-E)^2}{E}$	Rolling Chi-Squared Sum of Fit Statistics	Significance of Chi-Squared	Chi-Squared Critical Value at 5%	Null Hypothesis: Data is Poisson
0	0	0.985	0.985	0.985			
1	4	2.463	0.960	1.945			
2	4	3.078	0.276	2.221			
3	1	2.565	0.955	3.176			
4	0	1.603	1.603	4.779			
5 or more	3	1.306	2.198	6.977	13.7%	9.488	Accept

Total	30	over a	12	month period
Average	2.5			

However, now comes the contentious bit perhaps ... over what range of data do we make the comparison? The Poisson Distribution is unbounded to the right, meaning that very large numbers are theoretically possible, although extremely unlikely. This gives us a number of options:

1. Should we not test that we have an observed zero when theoretically we should expect a value that rounds to zero? In which case, at what point to we just stop; do we take the next expected zero and test that too? (... *We haven't got time to go on to infinity, have we?*)
2. Do we just ignore everything to the right of (i.e. larger than) the observed data? In this case, the sum of the fitted distribution will be less that the number of observations we actually have, and the result will include an inherent bias.
3. Do we have a catch-all last group that puts everything together than is bigger than a particular value? In which case, which value do we choose ... the largest observed value, or a value larger than this largest value? Would it make a difference?

We have already implied in our comments that the first two options are non-starters, so we have to select a value to use with option 3. Let's go with a popular choice in some texts of using the largest observed value as the catch-all 'greater than or equal to the largest value'. Firstly, we need to consider the mechanics of the calculations as shown in Table 7.7 to create the appropriate Chi-Squared Test:

•   The first three columns have been read across from the first, second and fifth columns of Table 7.6. The only difference is that the last row now calculates the theoretical number of Poisson Occurrences we might expect for five or more repairs arising per month. The last cell of the third column is calculated as the number of time periods multiplied by the probability of having greater than four repairs per month. We do this as **TimePeriod*(1-POISSON.DIST(***MaxRepairs-1*,

MeanRepairs, TRUE). This takes the complement of the cumulative probability, i.e. the total probability through to 100%, or 1

- The fourth column computes the Fit Statistic by taking the difference between the observed data point (O) and the theoretical data point (E), squaring it and dividing by the theoretical value (E)
- In the fifth column, the rolling Chi-Squared Statistic, is the incremental cumulative of fourth column. Each row cell represents the change in Chi-Squared Statistic as an extra sequential data point is added
- The significance of Chi-Squared in the sixth column, is the probability of getting that value or higher if the null hypothesis is true that the data follows a Poisson Distribution. It is calculated in Microsoft Excel using the Right-Tailed Chi-Squared function **CHISQ.DIST.RT(*Chi-Square Statistic, df*)** where *df* are the degrees of freedom given by the number of intervals included less two (*one for the Poisson Parameter and one for the pot*). In the case of the first cell in the list, corresponding to 5 repairs per month (six intervals including zero), the number of degrees of freedom is 4 (6 intervals -1 distribution parameter -1)
- In the seventh column, we have calculated the Critical Value of the Chi-Squared Distribution at the 5% Level using the right-tailed inverse of a Chi-Squared Distribution function **CHISQ.INV.RT(*significance, df*)**. The degrees of freedom, *df*, are calculated as above
- Finally, in the last column (eighth) we can determine whether to accept or reject the null hypothesis that the observed data follows a Poisson Distribution. We do this on the basis that the significance returned being less than or greater than the 5% level, or that the Chi-Squared Statistic is less than or greater than the Critical Value at 5% significance

Note: We really only need one of the sixth or seventh columns, not both. We have shown both here to demonstrate their equivalence; it's a matter of personal preference.

In this example we have a probability of some 13.7% of getting a value of 5 or more, so we would accept the Null Hypothesis that the data was Poisson Distributed.

However, we might have some sympathy with the argument of whether we should test that getting zero occurrences of the higher arisings per month is consistent with a Poisson Distribution, so instead (*just for the fun of it, perhaps*), what if we argued that we want the 'balancing' data point to represent '6 or more' data points with an observed arising of zero? Table 7.8 repeats the test ... and perhaps returns a surprising, and probably disappointing result.

What is even more bizarre is that if we were to change the 'balancing' probability to be geared around seven or more, we would accept the Null Hypothesis again as shown in Table 7.9!

We clearly have an issue here to be managed ... depending on what range of values we choose to test we get a different result. *Oh dear!* We may be wondering does this always occur? The answer is 'no'. Let's consider the same model with the same

Table 7.8 Calculating the Chi-Squared Test for Goodness of Fit (Long Hand) over a 12 Month Period (2)

Repairs per Month	Observed Occurrences (O)	Expected Poisson Occurrences (E)	Fit Statistic $\frac{(O-E)^2}{E}$	Rolling Chi-Squared Sum of Fit Statistics	Significance of Chi-Squared	Chi-Squared Critical Value at 5%	Null Hypothesis: Data is Poisson
0	0	0.985	0.985	0.985			
1	4	2.463	0.960	1.945			
2	4	3.078	0.276	2.221			
3	1	2.565	0.955	3.176			
4	0	1.603	1.603	4.779			
5	3	0.802	6.029	10.808			
6 or more	0	0.504	0.504	11.312	4.6%	11.070	Reject

Total	30	over a	12	month period
Average	2.5			

Table 7.9 Calculating the Chi-Squared Test for Goodness of Fit (Long Hand) over a 12 Month Period (3)

Repairs per Month	Observed Occurrences (O)	Expected Poisson Occurrences (E)	Fit Statistic $\frac{(O-E)^2}{E}$	Rolling Chi-Squared Sum of Fit Statistics	Significance of Chi-Squared	Chi-Squared Critical Value at 5%	Null Hypothesis: Data is Poisson
0	0	0.985	0.985	0.985			
1	4	2.463	0.960	1.945			
2	4	3.078	0.276	2.221			
3	1	2.565	0.955	3.176			
4	0	1.603	1.603	4.779			
5	3	0.802	6.029	10.808			
6	0	0.334	0.334	11.142			
7 or more	0	0.170	0.170	11.312	7.9%	12.592	Accept

Total	30	over a	12	month period
Average	2.5			

first 12 months and extend it to a 24 months window, and re-run the equivalent tests. In Tables 7.10 and 7.11 the only difference is that the sample mean we use as the basis of the Poisson Distribution has dropped from 2.5 to 2.083. In this case the Null Hypothesis should be accepted in both cases, there is a drop in the significance level but the result is not even marginal.

We would get a similar pair of consistent results if we extended the period to 48 months with a sample mean rate of 1.979 repairs per month.

What can we take from this?

- Statistics aren't perfect, they are a guide only and estimators should use them carefully especially where results are marginal. However, in principle, their use normally enhances TRACEability (Transparent, Repeatable, Appropriate, Credible and Experientially-based)

Table 7.10 Calculating the Chi-Squared Test for Goodness of Fit (Long Hand) over a 24 Month Period (1)

Repairs per Month	Observed Occurrences (O)	Expected Poisson Occurrences (E)	Fit Statistic $\frac{(O-E)^2}{E}$	Rolling Chi-Squared Sum of Fit Statistics	Significance of Chi-Squared	Chi-Squared Critical Value at 5%	Null Hypothesis: Data is Poisson
0	2	2.988	0.327	0.327			
1	8	6.226	0.506	0.833			
2	7	6.485	0.041	0.873			
3	3	4.504	0.502	1.375			
4	1	2.346	0.772	2.147			
5 or more	3	1.452	1.652	3.799	43.4%	9.488	Accept

Total	50		over a	24	month period
Average	2.083				

Table 7.11 Calculating the Chi-Squared Test for Goodness of Fit (Long Hand) over a 24 Month Period (2)

Repairs per Month	Observed Occurrences (O)	Expected Poisson Occurrences (E)	Fit Statistic $\frac{(O-E)^2}{E}$	Rolling Chi-Squared Sum of Fit Statistics	Significance of Chi-Squared	Chi-Squared Critical Value at 5%	Null Hypothesis: Data is Poisson
0	2	2.988	0.327	0.327			
1	8	6.226	0.506	0.833			
2	7	6.485	0.041	0.873			
3	3	4.504	0.502	1.375			
4	1	2.346	0.772	2.147			
5	3	0.977	4.186	6.333			
6 or more	0	0.474	0.474	6.808	23.5%	11.070	Accept

Total	50		over a	24	month period
Average	2.083				

- Statistics become more reliable as the volume of data used to generate them increases
- Overall, we would accept that this example does support the repair arrivals being a Poisson Process, which is very encouraging when we consider that the example was generated at random on that assumption. However, for a relatively small window, the statistics were not totally supportive and a judgement call would have had to be made if this was real data

Note: If we had inappropriately used Microsoft Excel's **CHISQ.TEST** function here, it would have led us to accept the null hypothesis when we tested the 'balancing probability' set at '5 or more', or as '6 or more'. However, in the spirit of TRACEability, we should conclude that:

It is better to make the wrong decision for the right reason rather than the right decision for the wrong reason?

The counter argument to this would be the 'ends justify the means' ... but that fails to deliver TRACEability.

However, there are instances where an assumption of a steady state Arising Rate is not appropriate, and we have to consider alternatives.

7.6 When is Weibull viable?

We may recall (from Volume II Chapter 4) the Weibull Distribution (*pronounced 'Vaybull'*) ... it was the one with the '*vibble-vobble* zone' that is used for Reliability Analysis over the life cycle of a product or project. Intuitively, it seems a reasonable distribution to consider here. To save us looking back to Volume II Chapter 4, let's remind ourselves of its key features:

- In its basic form, the Weibull Distribution is a 2-parameter distribution that models the change in reliability over the product life cycle. It can be used to model the probability of failure over time
- Microsoft Excel has a specific function for the distribution:

WEIBULL.DIST(*x, alpha, beta, cumulative*)

x is the elapsed time relative to a start time of zero; *alpha* is the shape parameter; *beta* is the scale parameter, and *cumulative* is the parameter that tells Excel whether we want the PDF or CDF probability curve

- When *alpha* < 1, the failure rate decreases over time, indicating 'infant mortality'
- When *alpha* > 1, the failure rate increases over time, indicating an ageing process (*I think my alpha is greater than one now; parts are beginning to fail more frequently than they used to do!*)
- When *alpha* = 1, the failure rate is constant over time, indicating perhaps that failures are due to random events

However, something special happens to the Weibull Distribution when *alpha* = 1 ...

- When *alpha* = 1, we have a special case of the Weibull Distribution ... we call it the **Exponential Distribution**. (Note: The Excel Function **WEIBULL.DIST** incorrectly returns a zero when x = 0, unlike the **EXPON.DIST** which returns the correct value of its scale parameter)
- Furthermore, its scale parameter is the reciprocal of the Weibull Scale parameter ... just as it was with the special case of the Gamma Distribution. (*it's funny how all these things tie themselves together*). See Figure 7.20

WEIBULL.DIST(*x, 1, beta, cumulative*) = EXPON.DIST
(x, 1/beta, cumulative)

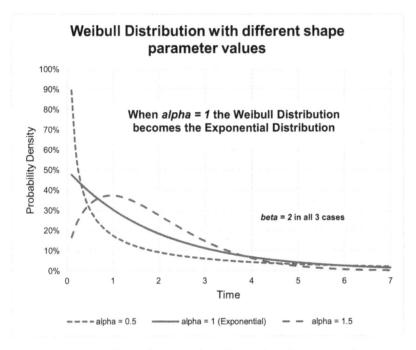

Figure 7.20 Weibull Distribution with Different Shape Parameter Values

The implication here is that a Poisson Process with its Exponential Inter-Arrival times is indicative of a Steady State failure rate ... so rather than being an assumption, we could argue that it is a natural consequence of a wider empirical relationship exhibited by the Weibull Distribution. However, this argument is a little contrived because the context in which we have used the two distributions is different!

Talking of Empirical Relationships ...

- A Weibull Distribution can also be used to model Production and Material Lead-times
- Often, we will have situations where failure rates increase or decrease over time
- We can also get 'Bath Tub' curves where failure rates decrease initially, then follow a steady state period before increasing again later in the product life (see Figure 7.21). These can be represented by more generic distributions such as the Exponentiated Weibull Distribution ... but these are not readily supported in Microsoft Excel, so we're not going to go there from here.

Given that these are empirical relationships, is there a practical alternative that allows us to *tap* into the Bath Tub Curve in Microsoft Excel? (*You didn't think that I would let that one go, did you?*) Cue for another queueing model ...

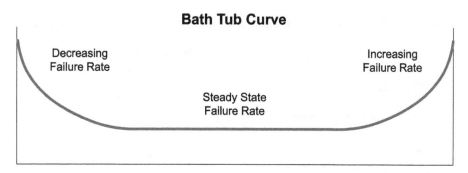

Figure 7.21 A Bath Tub Curve Generated by an Exponentiated Weibull Distribution

7.7 Can we have a Poisson Process with an increasing/ decreasing trend?

In Volume III Chapters 4 and 6 we discussed the natural scatter of data around a Linear or Logarithmic function as being Normally Distributed, whereas for an Exponential or Power function, the scatter was Lognormally Distributed, in other words, for the latter two the scatter is not symmetrical but positively skewed with more densely concentrated number of points below but close to the curve but more widely scattered above the line when viewed in linear space.

If we consider a line (or a curve for that matter) that increases or decreases over time, then we can conceive of the points being scattered around the line or curve as a Poisson Distribution. For example, as the number of aircraft of a particular type enter into service, then the demand for repairs will increase, but at any moment in time the number of such repair arisings per month can be assumed to be scattered as a Poisson Distribution. So, in answer to the question posed in the section title: yes.

If we already have data (as opposed to just an experientially or theoretically-based assumption) and we want to examine the underlying trend of the Mean of the data, then there are some techniques which are useful, ... and some that are not. We have summarised some of these in Table 7.12.

Let's create a model of what it might look like.

Suppose we have a situation where we have an initial repair arising rate of some four per month, falling exponentially over a 12-month period to two per month, with a rate of decline of 6% per month in terms of the average arising rate. After that let's assume a steady state rate of two per month. Table 7.13 illustrates such a set-up.

1. Cell C1 is the Initial Mean Arrival Rate and Cell C2 is the Exponential decrease factor per month

Table 7.12 Some Potential Techniques for Analysing Arising Rate Trends

Trend Analysis	Advantages		Disadvantages	
Moving Average	Simple. Tracks "central value"		No natural "moving" base – smoothing is by trial and error. Technique for trend smoothing only - not predictive	Sensitive to extreme values
Moving Median	Simple. Tracks "central value". Insensitive to extreme values			
Moving Minimum/ Maximum	Simple. Tracks lower and upper limits as crude Prediction Limits			Slow response times
Exponential Smoothing	Relatively simple. Provides for correction to previous forecasting errors		Very sensitive to volatile data	Choice of Damping Factor is subjective
Linear Regression	Gives the Line of Best Fit (LoBF)	Statistical measures for the goodness of fit	Often have to reject the LoBF with volatile arising data.	Negative trends will ultimately imply negative arisings
Exponential Regression	Gives the Exponential (Log-Linear) Curve of Best Fit (CoBF)		Often have to reject the CoBF with volatile arising data.	Will fail if we have zero arisings in any time period
Power Regression	Gives the Power (Log-Log) CoBF	Prediction Limits can be calculated		Never take the Log of a date!
Logarithmic Regression	Gives the Logarithmic (Linear-Log) CoBF			
Cumulative Smoothing	Naturally smoothes volatile arising data. Copes with zero arisings and extreme values		Can produce statistically acceptable but illogical results	

2. Cells C3 and J3 are the Average Steady State Repair Arrival Rate and Repair Rate respectively
3. Cells C4 and J4 are the Average Repair Inter-Arrival Time and the Average Repair Time expressed as the reciprocals of Cells C3 and J3 respectively. (Note that rows 3 and 4 are the opposite way round than we used for earlier models)
4. The process arrivals are generated by a series of random probabilities in Column B using **RAND()** in Microsoft Excel
5. Column C calculates the Time Adjusted Repair Arrival Rate using an Exponential decay based on the Initial Arrival Rate in Cell C1, reducing at a rate equal to Cell C2 until the Rate equals the Steady State Rate in Cell C3. In this example Cell C5 is given by the expression **MAX(\$C\$1*(1-\$C\$2)^F7,\$C\$3)** which is the maximum of the Steady State Rate and the Time Adjusted Rate. We will look at the calculation of the relevant month in Step 8

Table 7.13 Example of a G/M/1 Queueing Model with an Initially Decreasing Arising Rate

	A	B	C	D	E	F	G	H	I	J	K	L	M
1	Initial Mean Arrival Rate	4	per Month	l									
2	Arrival Rate Reduction	6%	per Month										
3	Mean Arrival Rate	2	per Month	λ				Mean Repair Rate >	2.50	per Month	μ		
4	Mean Inter-Arrival Time	0.50	Months	$1/\lambda$				Mean Repair Time >	0.40	Months	$1/\mu$		
5													
6	Repair Number	Random Probability	Time Adjusted Arrival Rate per Month	Months since last Repair Receipt	Arrival Time (Cum Months from Start)	Repair Receipt Month	Repair Start Time (Cum Months from Start)	Repair Start Month	Random Probability	Repair Time in Months	Completion Time (Cum Months from Start)	Repair Return Month	Queueing Time (Months)
7	Start >			0	0	0	0	0		0	0	0	0
8	1	58.77%	4.00	0.221	0.221	1	0.221	1	62.02%	0.387	0.609	1	0.000
9	2	86.92%	3.76	0.541	0.762	1	0.762	1	30.05%	0.143	0.905	1	0.000
10	3	39.57%	3.76	0.134	0.896	1	0.905	1	62.17%	0.389	1.294	2	0.009
11	4	8.11%	3.76	0.022	0.919	1	1.294	2	18.66%	0.083	1.377	2	0.375
12	5	25.45%	3.76	0.078	0.997	1	1.377	2	61.00%	0.377	1.754	2	0.380
13	6	17.14%	3.76	0.050	1.047	2	1.754	2	68.37%	0.460	2.214	3	0.707
14	7	38.99%	3.53	0.140	1.187	2	2.214	3	22.62%	0.103	2.317	3	1.027
15	8	36.07%	3.53	0.127	1.313	2	2.317	3	81.10%	0.666	2.983	3	1.003
16	9	27.93%	3.53	0.093	1.406	2	2.983	3	92.61%	1.042	4.025	5	1.577
17	10	34.72%	3.53	0.121	1.527	2	4.025	5	94.30%	1.146	5.171	6	2.498
18	11	4.09%	3.53	0.012	1.539	2	5.171	6	78.19%	0.609	5.780	6	3.632
19	12	47.57%	3.53	0.183	1.721	2	5.780	6	6.14%	0.025	5.805	6	4.059

6. Column D contains the random Poisson Arrivals for the Repairs which, as we discussed earlier, is equivalent to Exponentially distributed Inter-Arrival Times between the consecutive repairs and based on the random probabilities in Column B. Also as discussed previously, in Section 7.3, these inverse Exponential values are generated using the inverse of a Gamma Distribution with the alpha shape parameter set to one: **GAMMA.INV(RAND(),1,*TimeAdjustedMean*)** where the parameter value *TimeAdjustedMean* is that of the variable in Column C from Step 5. (Remember that the Exponential Distribution is a special case of the Gamma Distribution)

7. Column E is the running total of Column D to give the point in time that the repair is received

8. Column F expresses the Month in which the repair arrives This is the value we use to determine the Time for calculating the Time Adjusted Mean for the next Repair. (We could have used the unrounded value in Step 7, Column E – it's a matter of personal choice)

9. Column G determines the earliest Repair Start Time by calculating the Maximum of either the current Repair's arrival time from Column E or the previous Repair's Completion Time from Column K (which we will discuss in Step 12)

10. Column H is similar to Column E but relates to the month in which the repair process is deemed to have started

11. Column I is assumed here to be a random repair duration with an Exponential Distribution and is calculated as **GAMMA.INV(RAND(), 1, $H$3)**. We could use another distribution if we have the evidence to support it; this would make the model a M/G/1 queueing system

12. Column K is the time at which the repair is completed as is the sum of the Start Time (Column G) and the Repair Duration (Column J)

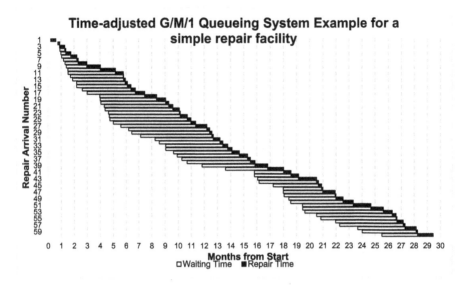

Figure 7.22 Example of Time-Adjusted G/M/1 Queueing Model

13. Column L is similar to Column F but relates to the month in which the Repair is deemed to have been completed

Figure 7.22 illustrates the output from this model.

We could build a multi-channel time-adjusted M/M/c or M/G/c Model by combining this technique with that which we discussed in Section 7.4 (*but to be honest, looking from here, it looks like many of us have had enough now, so we will not demonstrate it here*).

That's all we are going to cover in this introduction to Queueing Theory. (*I hope it was worth the wait.*)

7.8 Chapter review

Eventually (*sorry for making you wait so long*), we came to Queueing Theory and concentrated largely on how we might model different situations rather than manipulate parametric and algebraic formulae (*you should have seen the relief on your face!*), Perhaps the single most important learning point was that if the Demand Arisings into our system are distributed as a Poisson Distribution, then the Inter-Arrival Times between discrete demands will be Exponentially Distributed. (*Don't you love it when different models just tie themselves together consistently and neatly. No? Perhaps it's just me then.)*

In this journey into queueing we learnt that, as well as being spelt correctly, there are different Queueing Systems, which can be described by an internationally recognised coding convention, depicted by the notation A/S/c/K/N/D.

Many queueing scenarios rely on random arrivals, in which the time until the next arrival is independent of the time since the last arrival. This is property is referred to as being 'memoryless', and the Exponential is the only Continuous Probability Distribution with this property. We noted that a Geometric Distribution is the only Discrete Probability Distribution with this property.

We closed by considering the variation in demand over time by considering the viability of Weibull, used frequently to model failure rates (as well as manufacturing and process durations). We briefly considered the more complex end of the Weibull spectrum by acknowledging the empirical relationship of a Bath Tub Curve. However, we demonstrated how we could simulate this quite simplistically by splicing together Time-Adjusted Demand Model using Exponential Distributions, which, we noted, are special cases of the Weibull Distribution.

References

Department for Transport – Driving Standards Agency (2015) *The Official Highway Code*, London, The Stationary Office.

Erlang, AK (1909) 'The theory of probabilities and telephone conversations', *Nyt Tidsskrift for Matematik B*, Volume 20: pp.33–39.

Erlang, AK (1917) 'Solution of some problems in the theory of probabilities of significance in automatic telephone exchanges', *Elektrotkeknikeren*, Volume 13.

Kendall, DG (1953) 'Stochastic processes occurring in the theory of queues and their analysis by the method of the imbedded Markov Chain', *The Annals of Mathematical Statistics*, Volume 24, Number 3: p.338.

Kolmogorov, AN (1933) 'Grundbegriffe der Wahrscheinlichkeitrechnung', *Ergebnisse Der Mathematik*.

Little, JDC (1961) 'A proof for the Queuing Formula: L = λW', *Operations Research*, Volume 9, Number 3: pp.383–387.

Liu, HH (2009) *Software Performance and Scalability: A Quantitative Approach*, Hoboken, Wiley, pp. 361–362.

Mikes, G (1946) *How to Be an Alien*, London, Allan Wingate.

Stevenson, A & Waite, M (Eds) (2011) *Concise Oxford English Dictionary (12th Edition)*, Oxford, Oxford University Press

Wilkes, FM (1980) *Elements of Operational Research*, Maidenhead, McGraw-Hill.

Epilogue

Well, it took a long time getting here, but that was the final Chapter Review of this collection of numerical techniques for estimators and other number jugglers. Thank you for spending the time to read this offering and to work through the examples.

For me this has been a labour of love … some of us will already have commented that I need to get out more, perhaps now I can … or perhaps I should start researching all those things you wanted and expected to be in here, and regrettably have not been included. *Hmm, yes, perhaps I should start that tomorrow.*

Glossary of estimating and forecasting terms

This Glossary reflects those Estimating Terms that are either in common usage or have been defined for the purposes of this series of guides. Not all the terms are used in every volume, but where they do occur, their meaning is intended to be consistent.

3-Point Estimate   A 3-Point Estimate is an expression of uncertainty around an Estimate Value. It usually expresses Optimistic, Most Likely and Pessimistic Values.

Accuracy   Accuracy is an expression of how close a measurement, statistic or estimate is to the true value, or to a defined standard.

Actual Cost (AC)   See Earned Value Management Abbreviations and Terminology

ACWP (Actual Cost of Work Performed) or Actual Cost (AC)   See Earned Value Management Terminology.

Additive/Subtractive Time Series Model   See Time Series Analysis

Adjusted R-Square   Adjusted R-Square is a measure of the "Goodness of Fit" of a Multi-Linear Regression model to a set of data points, which reduces the Coefficient of Determination by a proportion of the Unexplained Variance relative to the Degrees of Freedom in the model, divided by the Degrees of Freedom in the Sum of Squares Error.

ADORE (Assumptions, Dependencies, Opportunities, Risks, Exclusions)   See Individual Terms.

Alternative Hypothesis   An Alternative Hypothesis is that supposition that the difference between an observed value and another observed or assumed value or effect, cannot be legitimately attributable to random sampling or experimental error. It is usually denoted as H_1.

Analogous Estimating Method or Analogy   See Analogical Estimating Method.

Analogical Estimating Method   The method of estimating by Analogy is a means of creating an estimate by comparing the similarities and/or differences between two things, one of which is used as the reference point against which rational adjustments for differences between the two things are made in order establish an estimate for the other.

Approach   See Estimating Approach.

Arithmetic Mean or Average   The Arithmetic Mean or Average of a set of numerical data values is a statistic calculated by summating the values of the individual terms and dividing by the number of terms in the set.

Assumption   An Assumption is something that we take to be broadly true or expect to come to fruition in the context of the Estimate.

Asymptote   An Asymptote to a given curve is a straight line that tends continually closer in value to that of the curve as they tend towards infinity (positive or negative). The difference between the asymptote and its curve reduces towards but never reaches zero at any finite value.

AT (Actual Time)   See Earned Value Management Abbreviations and Terminology

Average   See Arithmetic Mean.

Average (Mean) Absolute Deviation (AAD)   The Mean or Average Absolute Deviation of a range of data is the average 'absolute' distance of each data point from the Arithmetic Mean of all the data points, ignoring the sign depicting whether each point is less than or greater than the Arithmetic Mean.

Axiom   An Axiom is a statement or proposition that requires no proof, being generally accepted as being self-evidently true at all times.

BAC (Budget At Completion)   See Earned Value Management Abbreviations and Terminology.

Base Year Values   'Base Year Values' are values that have been adjusted to be expressed relative to a fixed year as a point of reference e.g., for contractual price agreement.

Basis of Estimate (BoE)   A Basis of Estimate is a series of statements that define the assumptions, dependencies and exclusions that bound the scope and validity of an estimate. A good BoE also defines the approach, method and potentially techniques used, as well as the source and value of key input variables, and as such supports Estimate TRACEability.

BCWP (Budgeted Cost of Work Performed)   See Earned Value Management Abbreviations and Terminology.

BCWS (Budgeted Cost of Work Scheduled)   See Earned Value Management Abbreviations and Terminology.

Benford's Law   Benford's Law is an empirical observation that in many situations the first or leading digit in a set of apparently random measurements follows a repeating pattern that can be predicted as the Logarithm of one plus the reciprocal of the leading digit. It is used predominately in the detection of fraud.

Bessel's Correction Factor   In general, the variance (and standard deviation) of a data sample will understate the variance (and standard deviation) of the underlying data population. Bessel's Correction Factor allows for an adjustment to be made so that the sample variance can be used as an unbiased estimator of the population variance. The adjustment requires that the Sum of Squares of the Deviations from the Sample Mean be divided one less than the number of observations or data points i.e. n-1 rather than the more intuitive the number of observations. Microsoft Excel takes this adjustment into account.

Bottom-up Approach   In a Bottom-up Approach to estimating, the estimator identifies the lowest level at which it is appropriate to create a range of estimates based on the task definition available, or that can be inferred. The overall estimate, or higher level summaries, typically through a Work Breakdown Structure, can be produced through incremental aggregation of the lower level estimates. A Bottom-up Approach requires a good definition of the task to be estimated, and is frequently referred to as detailed estimating or as engineering build-up.

Chauvenet's Criterion   A test for a single Outlier based on the deviation Z-Score of the suspect data point.

Chi-Squared Test or χ^2-Test   The Chi-Squared Test is a "goodness of fit" test that compares the variance of a sample against the variance of a theoretical or assumed distribution.

Classical Decomposition Method (Time Series)   Classical Decomposition Method is a means of analysing data for which there is a seasonal and/or cyclical pattern of variation. Typically, the underlying trend is identified, from which the average deviation or variation by season can be determined. The method can be used for multiplicative and additive/subtractive Time Series Models.

Closed Interval   A Closed Continuous Interval is one which includes its endpoints, and is usually depicted with square brackets: [Minimum, Maximum].

Coefficient of Determination   The Coefficient of Determination is a statistical index which measures how much of the total variance in one variable can be explained by the variance in the other variable. It provides a measure of how well the relationship between two variables can be represented by a straight line.

Coefficient of Variation (CV)   The Coefficient of Variation of a set of sample data values is a dimensionless statistic which expresses the ratio of the sample's Standard Deviation to its Arithmetic Mean. In the rare cases where the set of data is the entire population, then the Coefficient of Variation is expressed as the ratio of the population's Standard Deviation to its Arithmetic Mean. It can be expressed as either a decimal or percentage.

Collaborative Working   Collaborative Working is a term that refers to the management strategy of dividing a task between multiple partners working towards a common goal where there a project may be unviable for a single organisation. There is usually a cost penalty of such collaboration as it tends to create duplication in management and in integration activities.

Collinearity & Multicollinearity   Collinearity is an expression of the degree to which two supposedly independent predicator variables are correlated in the context of the observed values being used to model their relationship with the dependent variable that we wish to estimate. Multicollinearity is an expression to which collinearity can be observed across several predicator variables.

Complementary Cumulative Distribution Function (CCDF)   The Complementary Cumulative Distribution Function is the theoretical or observed probability of that variable being greater than a given value. It is calculated as the difference between 1 (or 100%) and the Cumulative Distribution Function, 1-CDF.

Composite Index   A Composite Index is one that has been created as the weighted average of a number of other distinct Indices for different commodities.

Concave Curve   A curve in which the direction of curvature appears to bend towards a viewpoint on the x-axis, similar to one that would be observed when viewing the inside of a circle or sphere.

Cone of Uncertainty   A generic term that refers to the empirical observation that the range of estimate uncertainty or accuracy improves through the life of a project. It is typified by its cone or funnel shape appearance.

Confidence Interval   A Confidence Interval is an expression of the percentage probability that data will lie between two distinct Confidence Levels, known as the Lower and Upper Confidence Limits, based on a known or assumed distribution of data from either a sample or an entire population.
See also Prediction Interval.

Confidence Level   A Confidence Level is an expression of the percentage probability that data selected at random from a known or assumed distribution of data (either a sample or an entire population), will be less than or equal to a particular value.

Confidence Limits   The Lower and Upper Confidence Limits are the respective Confidence Levels that bound a Confidence Interval, and are expressions of the two percentage probabilities that data will be less or equal to the values specified based on the known or assumed distribution of data in question from either a sample or an entire population. See also Confidence Interval.

Constant Year Values   'Constant Year Values' are values that have been adjusted to take account of historical or future inflationary effects or other changes, and are expressed in relation to the Current Year Values for any defined year. They are often referred to as 'Real Year Values'.

Continuous Probability Distribution   A mathematical expression of the relative theoretical probability of a random variable which can take on any value from a real number range. The range may be bounded or unbounded in either direction.

Convex Curve   A curve in which the direction of curvature appears to bend away from a viewpoint on the x-axis, similar to one that would be observed when viewing the outside of a circle or sphere.

Copula   A Copula is a Multivariate Probability Distribution based exclusively on a number Uniform Marginal Probability Distributions (one for each variable).

Correlation   Correlation is a statistical relationship in which the values of two or more variables exhibit a tendency to change in relationship with one other. These variables are said to be positively (or directly) correlated if the values tend to move in the same direction, and negatively (or inversely) correlated if they tend to move in opposite directions.

Cost Driver   See Estimate Drivers.

Covariance   The Covariance between a set of paired values is a measure of the extent to which the paired data values are scattered around the paired Arithmetic Means. It is the average of the product of each paired variable from its Arithmetic Mean.

CPI (Cost Performance Index)   See Earned Value Management Abbreviations and Terminology.

Crawford's Unit Learning Curve   A Crawford Unit Learning Curve is an empirical relationship that expresses the reduction in time or cost of each unit produced as a power function of the cumulative number units produced.

Critical Path   The Critical Path at a point in time depicts the string of dependent activities or tasks in a schedule for which there is no float or queuing time. As such the length of the Critical Path represents the quickest time that the schedule can be currently completed based on the current assumed activity durations.

Cross-Impact Analysis   A Cross-Impact Analysis is a qualitative technique used to identify the most significant variables in a system by considering the impact of each variable on the other variables.

Cumulative Average   A Point Cumulative Average is a single term value calculated as the average of the current and all previous consecutive recorded input values that have occurred in a natural sequence.

A Moving Cumulative Average, sometimes referred to as a Cumulative Moving Average, is an array (a series or range of ordered values) of successive Point Cumulative Average terms calculated from all previous consecutive recorded input values that have occurred in a natural sequence.

Cumulative Distribution Function (CDF)   The Cumulative Distribution Function of a Discrete Random Variable expresses the theoretical or observed probability of that variable being less than or equal to any given value. It equates to the sum of the probabilities of achieving that value and each successive lower value.

The Cumulative Distribution Function of a Continuous Random Variable expresses the theoretical or observed probability of that variable being less than or equal to any given value. It equates to the area under the Probability Density Function curve to the left of the value in question.

See also the Complementary Cumulative Distribution Function.

Current Year (or Nominal Year) Values 'Current Year Values' are historical values expressed in terms of those that were current at the historical time at which they were incurred. In some cases, these may be referred to as 'Nominal Year Values'.

CV (Cost Variance) See Earned Value Management Abbreviations and Terminology

Data Type Primary Data is that which has been taken directly from its source, either directly or indirectly, without any adjustment to its values or context.

Secondary Data is that which has been taken from a known source, but has been subjected to some form of adjustment to its values or context, the general nature of which is known and has been considered to be appropriate.

Tertiary Data is data of unknown provenance. The specific source of data and its context is unknown, and it is likely that one or more adjustments of an unknown nature have been made, in order to make it suitable for public distribution.

Data Normalisation Data Normalisation is the act of making adjustments to, or categorisations of, data to achieve a state where data the can be used for comparative purposes in estimating.

Decile A Decile is one of ten subsets from a set of ordered values which nominally contain a tenth of the total number of values in each subset. The term can also be used to express the values that divide the ordered values into the ten ordered subsets.

Degrees of Freedom Degrees of Freedom are the number of different factors in a system or calculation of a statistic that can vary independently.

DeJong Unit Learning Curve A DeJong Unit Learning Curve is a variation of the Crawford Unit Learning Curve that allows for an incompressible or 'unlearnable' element of the task, expressed as a fixed cost or time.

Delphi Technique The Delphi Technique is a qualitative technique that promotes consensus or convergence of opinions to be achieved between diverse subject matter experts in the absence of a clear definition of a task or a lack of tangible evidence.

Dependency A Dependency is something to which an estimate is tied, usually an uncertain event outside of our control or influence, which if it were not to occur, would potentially render the estimated value invalid. If it is an internal dependency, the estimate and schedule should reflect this relationship

Descriptive Statistic A Descriptive Statistic is one which reports an indisputable and repeatable fact, based on the population or sample in question, and the nature of which is described in the name of the Statistic.

Discount Rate The Discount Rate is the percentage reduction used to calculate the present-day values of future cash flows. The discount rate often either reflects the comparable market return on investment of opportunities with similar levels of risk, or reflects an organisation's Weighted Average Cost of Capital (WACC), which is based on the weighted average of interest rates paid on debt (loans) and shareholders' return on equity investment.

Discounted Cash Flow (DCF) Discounted Cash Flow (DCF) is a technique for converting estimated or actual expenditures and revenues to economically comparable values at a common point in time by discounting future cash flows by an agreed percentage discount rate per time period, based on the cost to the organisation of borrowing money, or the average return on comparable investments.

Discrete Probability Distribution A mathematical expression of the theoretical or empirical probability of a random variable which can only take on predefined values from a finite range.

Dixon's Q-Test A test for a single Outlier based on the distance between the suspect data point and its nearest neighbour in comparison with the overall range of the data.

Driver See Estimate Drivers.

Earned Value (EV) See Earned Value Management Terminology.

Earned Value Management (EVM) Earned Value Management is a collective term for the management and control of project scope, schedule and cost.

Earned Value Analysis Earned Value Analysis is a collective term used to refer to the analysis of data gathered and used in an Earned Value Management environment.

Earned Value Management Abbreviations and Terminology (Selected terms only)

ACWP (Actual Cost of Work Performed) sometimes referred to as Actual Cost (AC) Each point represents the cumulative actual cost of the work completed or in progress at that point in time. The curve represents the profile by which the actual cost has been expended for the value achieved over time.

AT (Actual Time) AT measures the time from start to time now.

BAC (Budget At Completion) The BAC refers to the agreed target value for the current scope of work, against which overall performance will be assessed.

BCWP (Budget Cost of Work Performed) sometimes referred to as Earned Value (EV) Each point represents the cumulative budgeted cost of the work completed or in progress to that point in time. The curve represents the profile by which the budgeted cost has been expended over time. The BCWP is expressed in relation to the BAC (Budget At Completion).

BCWS sometimes referred to as Planned Value (PV) Each point represents the cumulative budgeted cost of the work planned to be completed or to be in progress to that point in time. The curve represents the profile by which the budgeted cost was planned to be expended over time. The BCWS is expressed in relation to the BAC (Budget At Completion).

CPI (Cost Performance Index) The CPI is an expression of the relative performance from a cost perspective and is the ratio of Earned Value to Actual Cost (EV/AC) or (BCWP/ACWP).

CV (Cost Variance) CV is a measure of the cumulative Cost Variance as the difference between the Earned Value and the Actual Cost (EV − AC) or (BCWP − ACWP).

ES (Earned Schedule) ES measures the planned time allowed to reach the point that we have currently achieved.

EAC (Estimate At Completion) sometimes referred to as FAC (Forecast At Completion) The EAC or FAC is the sum of the actual cost to date for the work achieved, plus an estimate of the cost to complete any outstanding or incomplete activity or task in the defined scope of work

ETC (Estimate To Completion) The ETC is an estimate of the cost that is likely to be expended on the remaining tasks to complete the current scope of agreed work. It is the difference between the Estimate At Completion and the current Actual Cost (EAC − ACWP or AC).

SPI (Schedule Performance Index)   The SPI is an expression of the relative schedule performance expressed from a cost perspective and is the ratio of Earned Value to Planned Value (EV/PV) or (BCWP/BCWS). It is now considered to be an inferior measure of true schedule variance in comparison with SPI(t).

SPI(t)   The SPI(t) is an expression of the relative schedule performance and is the ratio of Earned Schedule to Actual Time (ES/AT).

SV (Schedule Variance)   SV is a measure of the cumulative Schedule Variance measured from a Cost Variance perspective, and is the difference between the Earned Value and the Planned Value (EV − PV) or (BCWP − BCWS). It is now considered to be an inferior measure of true schedule variance in comparison with SV(t).

SV(t)   SV(t) is a measure of the cumulative Schedule Variance and is the difference between the Earned Schedule and the Actual Time (ES − AT).

Equivalent Unit Learning   Equivalent Unit Learning is a technique that can be applied to complex programmes of recurring activities to take account of Work-in-Progress and can be used to give an early warning indicator of potential learning curve breakpoints. It can be used to supplement traditional completed Unit Learning Curve monitoring.

ES (Earned Schedule)   See Earned Value Management Abbreviations and Terminology

Estimate   An Estimate for 'something' is a numerical expression of the approximate value that might reasonably be expected to occur based on a given context, which is described and is bounded by a number of parameters and assumptions, all of which are pertinent to and necessarily accompany the numerical value provided.

Estimate At Completion (EAC) and **Estimate To Completion (ETC)**   See Earned Value Management Abbreviations and Terminology.

Estimate Drivers   A Primary Driver is a technical, physical, programmatic or transactional characteristic that either causes a major change in the value being estimated or in a major constituent element of it, or whose value itself changes correspondingly with the value being estimated, and therefore, can be used as an indicator of a change in that value.

A Secondary Driver is a technical, physical, programmatic or transactional characteristic that either causes a minor change in the value being estimated or in a constituent element of it, or whose value itself changes correspondingly with the value being estimated and can be used as an indicator of a subtle change in that value.

Cost Drivers are specific Estimate Drivers that relate to an indication of Cost behaviour.

Estimate Maturity Assessment (EMA)   An Estimate Maturity Assessment provides a 'health warning' on the maturity of an estimate based on its Basis of Estimate, and takes account of the level of task definition available and historical evidence used.

Estimating Approach   An Estimating Approach describes the direction by which the lowest level of detail to be estimated is determined.

See also Bottom-up Approach, Top-down Approach and Ethereal Approach.

Estimating Method   An Estimating Method is a systematic means of creating an estimate, or an element of an estimate. An Estimating Methodology is a set or system of Estimating Methods.

See also Analogous Method, Parametric Method and Trusted Source Method.

Estimating Metric   An Estimating Metric is a value or statistic that expresses a numerical relationship between a value for which an estimate is required, and a Primary or Secondary Driver (or parameter) of that value, or in relation to some fixed reference point.

See also Factor, Rate and Ratio.

Estimating Procedure An Estimating Procedure is a series of steps conducted in a certain manner and sequence to optimise the output of an Estimating Approach, Method and/or Technique.

Estimating Process An Estimating Process is a series of mandatory or possibly optional actions or steps taken within an organisation, usually in a defined sequence or order, in order to plan, generate and approve an estimate for a specific business purpose.

Estimating Technique An Estimating Technique is a series of actions or steps conducted in an efficient manner to achieve a specific purpose as part of a wider Estimating Method. Techniques can be qualitative as well as quantitative.

Ethereal Approach An Ethereal Approach to Estimating is one in which values are accepted into the estimating process, the provenance of which is unknown and at best may be assumed. These are values often created by an external source for low value elements of work, or by other organisations with acknowledged expertise. Other values may be generated by Subject Matter Experts internal to the organisation where there is insufficient definition or data to produce an estimate by a more analytical approach.

The Ethereal Approach should be considered the approach of last resort where low maturity is considered acceptable. The approach should be reserved for low value elements or work, and situations where a robust estimate is not considered critical.

Excess Kurtosis The Excess Kurtosis is an expression of the relative degree of Peakedness or flatness of a set of data values, relative to a Normal Distribution. Flatter distributions with a negative Excess Kurtosis are referred to as Platykurtic; Peakier distributions with a positive Excess Kurtosis are termed Leptokurtic; whereas those similar to a Normal Distribution are said to be Mesokurtic. The measure is based on the fourth power of the deviation around the Arithmetic Mean.

Exclusion An Exclusion is condition or set of circumstances that have been designated to be out of scope of the current estimating activities and their output.

Exponential Function An Exponential Function of two variables is one in which the Logarithm of the dependent variable on the vertical axis produces a monotonic increasing or decreasing Straight Line when plotted against the independent variable on the horizontal axis.

Exponential Smoothing Exponential Smoothing is a 'single-point' predictive technique which generates a forecast for any period based on the forecast made for the prior period, adjusted for the error in that prior period's forecast.

Extrapolation The act of estimating a value extrinsic to or outside the range of the data being used to determine that value. See also Interpolation.

Factored or Expected Value Technique A technique that expresses an estimate based on the weighted sum of all possible values multiplied by the probability of arising.

Factors, Rates and Ratios See individual terms: Factor Metric, Rate Metric and Ratio Metric

Factor Metric A Factor is an Estimating Metric used to express one variable's value as a percentage of another variable's value.

F-Test The F-Test is a "goodness of fit" test that returns the cumulative probability of getting an F-Statistic less than or equal to the ratio inferred by the variances in two samples.

Generalised Exponential Function A variation to the standard Exponential Function which allows for a constant value to exist in the dependent or predicted variable's value. It effectively creates a vertical shift in comparison with a standard Exponential Function.

Generalised Extreme Studentised Deviate A test for multiple Outliers based on the deviation Z-Score of the suspect data point.

Generalised Logarithmic Function   A variation to the standard Logarithmic Function which allows for a constant value to exist in the independent or predictor variable's value. It effectively creates a horizontal shift in comparison with a standard Logarithmic Function.

Generalised Power Function   A variation to the standard Power Function which allows for a constant value to exist in either or both the independent and dependent variables' value. It effectively creates a horizontal and/or vertical shift in comparison with a standard Power Function.

Geometric Mean   The Geometric Mean of a set of n numerical data values is a statistic calculated by taking the nth root of the product of the n terms in the set.

Good Practice Spreadsheet Modelling (GPSM)   Good Practice Spreadsheet Modelling Principles relate to those recommended practices that should be considered when developing a Spreadsheet in order to help maintain its integrity and reduce the risk of current and future errors.

Grubbs' Test   A test for a single Outlier based on the deviation Z-Score of the suspect data point.

Harmonic Mean   The Harmonic Mean of a set of n numerical data values is a statistic calculated by taking the reciprocal of the Arithmetic Mean of the reciprocals of the n terms in the set.

Heteroscedasticity   Data is said to exhibit Heteroscedasticity if data variances are not equal for all data values.

Homoscedasticity   Data is said to exhibit Homoscedasticity if data variances are equal for all data values.

Iglewicz and Hoaglin's M-Score (Modified Z-Score)   A test for a single Outlier based on the Median Absolute Deviation of the suspect data point.

Index   An index is an empirical average factor used to increase or decrease a known reference value to take account of cumulative changes in the environment, or observed circumstances, over a period of time. Indices are often used as to normalise data.

Inferential Statistic   An Inferential Statistic is one which infers something, often about the wider data population, based on one or more Descriptive Statistics for a sample, and as such, it is open to interpretation ... and disagreement.

Inherent Risk in Spreadsheets (IRiS)   IRiS is a qualitative assessment tool that can be used to assess the inherent risk in spreadsheets by not following Good Practice Spreadsheets Principles.

Interdecile Range   The Interdecile Range comprises the middle eight Decile ranges and represents the 80% Confidence Interval between the 10% and 90% Confidence Levels for the data.

Internal Rate of Return   The Internal Rate of Return (IRR) of an investment is that Discount Rate which returns a Net Present Value (NPV) of zero, i.e. the investment breaks even over its life with no over or under recovery.

Interpolation   The act of estimating an intermediary or intrinsic value within the range of the data being used to determine that value. See also Extrapolation.

Interquantile Range   An Interquantile Range is a generic term for the group of Quantiles that form a symmetrical Confidence Interval around the Median by excluding the first and last Quantile ranges.

Interquartile Range   The Interquartile Range comprises the middle two Quartile ranges and represents the 50% Confidence Interval between the 25% and 75% Confidence Levels for the data.

Interquintile Range   The Interquintile Range comprises the middle three Quintile ranges and represents the 60% Confidence Interval between the 20% and 80% Confidence Levels for the data.

Jarque-Bera Test   The Jarque-Bera Test is a statistical test for whether data can be assumed to follow a Normal Distribution. It exploits the properties of a Normal Distribution's Skewness and Excess Kurtosis being zero.

Kendall's Tau Rank Correlation Coefficient   Kendall's Tau Rank Correlation Coefficient for two variables is a statistic that measures the difference between the number of Concordant and Discordant data pairs as a proportion of the total number of possible unique pairings, where two pairs are said to be concordant if the ranks of the two variables move in the same direction, or are said to be discordant if the ranks of the two variables move in opposite directions.

Laspeyres Index   Laspeyres Indices are time-based indices which compare the prices of commodities at a point in time with the equivalent prices for the Index Base Period, based on the original quantities consumed at the Index Base Year.

Learning Curve   A Learning Curve is a mathematical representation of the degree at which the cost, time or effort to perform one or more activities reduces through the acquisition and application of knowledge and experience through repetition and practice.

Learning Curve Breakpoint   A Learning Curve Breakpoint is the position in the build or repetition sequence at which the empirical or theoretical rate of learning changes.

Learning Curve Cost Driver   A Learning Curve Cost Driver is an independent variable which affects or indicates the rate or amount of learning observed.

Learning Curve Segmentation   Learning Curve Segmentation refers to a technique which models the impact of discrete Learning Curve Cost Drivers as a product of multiple unit-based learning curves.

Learning Curve Step-point   A Learning Curve Step-point is the position in the build or repetition sequence at which there is a step function increase or decrease in the level of values evident on the empirical or theoretical Learning Curve.

Learning Exponent   A Learning Exponent is the power function exponent of a Learning Curve reduction and is calculated as the Logarithmic value of the Learning Rate using a Logarithmic Base equivalent to the Learning Rate Multiplier.

Learning Rate and Learning Rate Multiplier   The Learning Rate expresses the complement of the percentage reduction over a given Learning Rate Multiplier (usually 2). For example, an 80% Learning Rate with a Learning Multiplier of 2 implies a 20% reduction every time the quantity doubles.

Least Squares Regression   Least Squares Regression is a Regression procedure which identifies the 'Best Fit' of a pre-defined functional form by minimising the Sum of the Squares of the vertical difference between each data observation and the assumed functional form through the Arithmetic Mean of the data.

Leptokurtotic or Leptokurtic   An expression that the degree of Excess Kurtosis in a probability distribution is peakier than a Normal Distribution.

Linear Function   A Linear Function of two variables is one which can be represented as a monotonic increasing or decreasing Straight Line without any need for Mathematical Transformation.

Logarithm   The Logarithm of any positive value for a given positive Base Number not equal to one is that power to which the Base Number must be raised to get the value in question.

Logarithmic Function A Logarithmic Function of two variables is one in which the dependent variable on the vertical axis produces a monotonic increasing or decreasing Straight Line, when plotted against the Logarithm of the independent variable on the horizontal axis.

Mann–Whitney U-Test sometimes known as Mann-Whitney-Wilcoxon U-Test A U-Test is used to test whether two samples could be drawn from the same population by comparing the distribution of the joint ranks across the two samples.

Marching Army Technique sometimes referred to as Standing Army Technique The Marching Army Technique refers to a technique that assumes that costs vary directly in proportion with a schedule.

Mathematical Transformation A Mathematical Transformation is a numerical process in which the form, nature or appearance of a numerical expression is converted into an equivalent but non-identical numerical expression with a different form, nature or appearance.

Maximum The Maximum is the largest observed value in a sample of data, or the largest potential value in a known or assumed statistical distribution. In some circumstances, the term may be used to imply a pessimistic value at the upper end of potential values rather than an absolute value.

Mean Absolute Deviation See Average Absolute Deviation (AAD).

Measures of Central Tendency Measures of Central Tendency is a collective term that refers to those descriptive statistics that measure key attributes of a data sample (Means, Modes and Median).

Measures of Dispersion and Shape Measures of Dispersion and Shape is a collective term that refers to those descriptive statistics that measure the degree and/or pattern of scatter in the data in relation to the Measures of Central Tendency.

Median The Median of a set of data is that value which occurs in the middle of the sequence when its values have been arranged in ascending or descending order. There are an equal number of data points less than and greater than the Median.

Median Absolute Deviation (MAD) The Median Absolute Deviation of a range of data is the Median of the 'absolute' distance of each data point from the Median of those data points, ignoring the "sign" depicting whether each point is less than or greater than the Median.

Memoryless Probability Distribution In relation to Queueing Theory, a Memoryless Probability Distribution is one in which the probability of waiting a set period of time is independent of how long we have been waiting already. The probability of waiting longer than the sum of two values is the product of the probabilities of waiting longer than each value in turn. An Exponential Distribution is the only Continuous Probability Distribution that exhibits this property, and a Geometric Distribution is the only discrete form.

Mesokurtotic or Mesokurtic An expression that the degree of Excess Kurtosis in a probability distribution is comparable with a Normal Distribution.

Method See Estimating Method.

Metric A Metric is a statistic that measures an output of a process or a relationship between a variable and another variable or some reference point.
See also Estimating Metric.

Minimum The Minimum is the smallest observed value in a sample of data, or the smallest potential value in a known or assumed statistical distribution. In some circumstances, the term may be used to imply an optimistic value at the lower end of potential values rather than an absolute value.

Mode   The Mode of a set of data is that value which has occurred most frequently, or that which has the greatest probability of occurring.

Model Validation and Verification   See individual terms: Validation and Verification.

Monotonic Function   A Monotonic Function of two paired variables is one that when values are arranged in ascending numerical order of one variable, the value of the other variable either perpetually increases or perpetually decreases.

Monte Carlo Simulation   Monte Carlo Simulation is a technique that models the range and relative probabilities of occurrence, of the potential outcomes of a number of input variables whose values are uncertain but can be defined as probability distributions.

Moving Average   A Moving Average is a series or sequence of successive averages calculated from a fixed number of consecutive input values that have occurred in a natural sequence. The fixed number of consecutive input terms used to calculate each average term is referred to as the Moving Average Interval or Base.

Moving Geometric Mean   A Moving Geometric Mean is a series or sequence of successive geometric means calculated from a fixed number of consecutive input values that have occurred in a natural sequence. The fixed number of consecutive input terms used to calculate each geometric mean term is referred to as the Moving Geometric Mean Interval or Base.

Moving Harmonic Mean   A Moving Harmonic Mean is a series or sequence of successive harmonic means calculated from a fixed number of consecutive input values that have occurred in a natural sequence. The fixed number of consecutive input terms used to calculate each harmonic mean term is referred to as the Moving Harmonic Mean Interval or Base.

Moving Maximum   A Moving Maximum is a series or sequence of successive maxima calculated from a fixed number of consecutive input values that have occurred in a natural sequence. The fixed number of consecutive input terms used to calculate each maximum term is referred to as the Moving Maximum Interval or Base.

Moving Median   A Moving Median is a series or sequence of successive medians calculated from a fixed number of consecutive input values that have occurred in a natural sequence. The fixed number of consecutive input terms used to calculate each median term is referred to as the Moving Median Interval or Base.

Moving Minimum   A Moving Minimum is a series or sequence of successive minima calculated from a fixed number of consecutive input values that have occurred in a natural sequence. The fixed number of consecutive input terms used to calculate each minimum term is referred to as the Moving Minimum Interval or Base.

Moving Standard Deviation   A Moving Standard Deviation is a series or sequence of successive standard deviations calculated from a fixed number of consecutive input values that have occurred in a natural sequence. The fixed number of consecutive input terms used to calculate each standard deviation term is referred to as the Moving Standard Deviation Interval or Base.

Multicollinearity   See Collinearity.

Multiplicative Time Series Model   See Time Series Analysis.

Multi-Variant Unit Learning   Multi-Variant Unit Learning is a technique that considers shared and unique learning across multiple variants of the same or similar recurring products.

Net Present Value   The Net Present Value (NPV) of an investment is the sum of all positive and negative cash flows through time, each of which have been discounted based on the time value of money relative to a Base Year (usually the present year).

Nominal Year Values   'Nominal Year Values' are historical values expressed in terms of those that were current at the historical time at which they were incurred. In some cases, these may be referred to as 'Current Year Values'.

Norden-Rayleigh Curve   A Norden-Rayleigh is an empirical relationship that models the distribution of resource required in the non-recurring concept demonstration or design and development phases.

Null Hypothesis   A Null Hypothesis is that supposition that the difference between an observed value or effect and another observed or assumed value or effect, can be legitimately attributable to random sampling or experimental error. It is usually denoted as H_0.

Open Interval   An Open Continuous Interval is one which excludes its endpoints, and is usually depicted with rounded brackets: (Minimum, Maximum).

Opportunity   An Opportunity is an event or set of circumstances that may or may not occur, but if it does occur an Opportunity will have a beneficial effect on our plans, impacting positively on the cost, quality, schedule, scope compliance and/or reputation of our project or organisation.

Optimism Bias   Optimism Bias is an expression of the inherent bias (often unintended) in an estimate output based on either incomplete or misunderstood input assumptions.

Outlier   An outlier is a value that falls substantially outside the pattern of other data. The outlier may be representative of unintended atypical factors or may simply be a value which has a very low probability of occurrence.

Outturn Year Values   'Outturn Year Values' are values that have been adjusted to express an expectation of what might be incurred in the future due to escalation or other predicted changes. In some cases, these may be referred to as 'Then Year Values'.

Paasche Index   Paasche Indices are time-based indices which compare prices of commodities at a point in time with the equivalent prices for the Index Base Period, based on the quantities consumed at the current point in time in question

Parametric Estimating Method   A Parametric Estimating Method is a systematic means of establishing and exploiting a pattern of behaviour between the variable that we want to estimate, and some other independent variable or set of variables or characteristics that have an influence on its value.

Payback Period   The Payback Period is an expression of how long it takes for an investment opportunity to break even, i.e. to pay back the investment.

Pearson's Linear Correlation Coefficient   Pearson's Linear Correlation Coefficient for two variables is a measure of the extent to which a change in the value of one variable can be associated with a change in the value of the other variable through a linear relationship. As such it is a measure of linear dependence or linearity between the two variables, and can be calculated by dividing the Covariance of the two variables by the Standard Deviation of each variable.

Peirce's Criterion   A test for multiple Outliers based on the deviation Z-Score of the suspect data point.

Percentile   A Percentile is one of a hundred subsets from a set of ordered values which each nominally contain a hundredth of the total number of values in each subset. The term can also be used to express the values that divide the ordered values into the hundred ordered subsets.

Planned Value (PV)   See Earned Value Management Abbreviations and Terminology.

Platykurtotic or Platykurtic   An expression that the degree of Excess Kurtosis in a probability distribution is shallower than a Normal Distribution.

Power Function   A Power Function of two variables is one in which the Logarithm of the dependent variable on the vertical axis produces a monotonic increasing or decreasing Straight Line when plotted against the Logarithm of the independent variable on the horizontal axis.

Precision
  (1) Precision is an expression of how close repeated trials or measurements are to each other.
  (2) Precision is an expression of the level of exactness reported in a measurement, statistic or estimate.

Primary Data   See Data Type.

Primary Driver   See Estimate Drivers.

Probability Density Function (PDF)   The Probability Density Function of a Continuous Random Variable expresses the rate of change in the probability distribution over the range of potential continuous values defined, and expresses the relative likelihood of getting one value in comparison with another.

Probability Mass Function (PMF)   The Probability Mass Function of a Discrete Random Variable expresses the probability of the variable being equal to each specific value in the range of all potential discrete values defined. The sum of these probabilities over all possible values equals 100%.

Probability of Occurrence   A Probability of Occurrence is a quantification of the likelihood that an associated Risk or Opportunity will occur with its consequential effects.

Quadratic Mean or Root Mean Square   The Quadratic Mean of a set of n numerical data values is a statistic calculated by taking the square root of the Arithmetic Mean of the squares of the n values. As a consequence, it is often referred to as the Root Mean Square.

Quantile   A Quantile is the generic term for a number of specific measures that divide a set of ordered values into a quantity of ranges with an equal proportion of the total number of values in each range. The term can also be used to express the values that divide the ordered values into such ranges.

Quantity-based Learning Curve   A Quantity-based Learning Curve is an empirical relationship which reflects that the time, effort or cost to perform an activity reduces as the number of repetitions of that activity increases.

Quartile   A Quartile is one of four subsets from a set of ordered values which nominally contain a quarter of the total number of values in each subset. The term can also be used to express the values that divide the ordered values into the four ordered subsets.

Queueing Theory   Queueing Theory is that branch of Operation Research that studies the formation and management of queuing systems and waiting times.

Quintile   A Quintile is one of five subsets from a set of ordered values which nominally contain a fifth of the total number of values in each subset. The term can also be used to express the values that divide the ordered values into the five ordered subsets.

Range   The Range is the difference between the Maximum and Minimum observed values in a dataset, or the Maximum and Minimum theoretical values in a statistical distribution. In some circumstances, the term may be used to imply the difference between pessimistic and optimistic values from the range of potential values rather than an absolute range value.

Rate Metric   A Rate is an Estimating Metric used to quantify how one variable's value changes in relation to some measurable driver, attribute or parameter, and would be expressed in the form of a [Value] of one attribute per [Unit] of another attribute.

Ratio Metric   A Ratio is an Estimating Metric used to quantify the relative size proportions between two different instances of the same driver, attribute or characteristic such as weight. It is typically used as an element of Estimating by Analogy or in the Normalisation of data.

Real Year Values   'Real Year Values' are values that have been adjusted to take account of historical or future inflationary effects or other changes, and are expressed in relation to the Current Year Values for any defined year. They are often referred to as 'Constant Year Values'.

Regression Analysis   Regression Analysis is a systematic procedure for establishing the Best Fit relationship of a predefined form between two or more variables, according to a set of Best Fit criteria.

Regression Confidence Interval   The Regression Confidence Interval of a given probability is an expression of the Uncertainty Range around the Regression Line. For a known value of a single independent variable, or a known combination of values from multiple independent variables, the mean of all future values of the dependent variable will occur within the Confidence Interval with the probability specified.

Regression Prediction Interval   A Regression Prediction Interval of a given probability is an expression of the Uncertainty Range around future values of the dependent variable based on the regression data available. For a known value of a single independent variable, or a known combination of values from multiple independent variables, the future value of the dependent variable will occur within the Prediction Interval with the probability specified.

Residual Risk Exposure   The Residual Risk Exposure is the weighted value of the Risk, calculated by multiplying its Most Likely Value by the complement of its Probability of Occurrence (100% − Probability of Occurrence). It is used to highlight the relative value of the risk that is not covered by Risk Exposure calculation.

Risk   A Risk is an event or set of circumstances that may or may not occur, but if it does occur a Risk will have a detrimental effect on our plans, impacting negatively on the cost, quality, schedule, scope compliance and/or reputation of our project or organisation.

Risk Exposure   A Risk Exposure is the weighted value of the Risk, calculated by multiplying its Most Likely Value by its Probability of Occurrence.
See also Residual Risk Exposure.

Risk & Opportunity Ranking Factor   A Risk & Opportunity Ranking Factor is the relative absolute exposure of a Risk or Opportunity in relation to all others, calculated by dividing the absolute value of the Risk Exposure by the sum of the absolute values of all such Risk Exposures.

Risk Uplift Factors   A Top-down Approach to Risk Analysis may utilise Risk Uplift Factors to quantify the potential level of risk based on either known risk exposure for the type of work being undertaken based on historical records of similar projects, or based on a Subject Matter Expert's Judgement.

R-Square (Regression)   R-Square is a measure of the "Goodness of Fit" of a simple linear regression model to a set of data points. It is directly equivalent to the Coefficient of Determination that shows how much of the total variance in one variable can be explained by the variance in the other variable.
See also Adjusted R-Square.

Schedule Maturity Assessment (SMA)   A Schedule Maturity Assessment provides a 'health warning' on the maturity of a schedule based on its underpinning assumptions and interdependencies, and takes account of the level of task definition available and historical evidence used.

Secondary Data   See Data Type.

Secondary Driver   See Estimate Drivers.

Skewness Coefficient   The Fisher-Pearson Skewness Coefficient is an expression of the degree of asymmetry of a set of values around their Arithmetic Mean. A positive Skewness Coefficient indicates that the data has a longer tail on the right-hand side, in the direction of the positive axis; such data is said to be Right or Positively Skewed. A negative Skewness Coefficient indicates that the data has a longer tail on the left-hand side, in the direction of the negative axis; such data is said to be Left or Negatively Skewed. Data that is distributed symmetrically returns a Skewness Coefficient of zero.

Slipping and Sliding Technique   A technique that compares and contrasts a Bottom-up Monte Carlo Simulation Cost evaluation of Risk, Opportunity and Uncertainty with a holistic Top-down Approach based on Schedule Risk Analysis and Uplift Factors.

Spearman's Rank Correlation Coefficient   Spearman's Rank Correlation Coefficient for two variables is a measure of monotonicity of the ranks of the two variables, i.e. the degree to which the ranks move in the same or opposite directions consistently. As such it is a measure of linear or non-linear interdependence.

SPI (Schedule Performance Index – Cost Impact)   See Earned Value Management Abbreviations and Terminology.

SPI(t) (Schedule Performance Index – Time Impact)   See Earned Value Management Abbreviations and Terminology.

Spreadsheet Validation and Verification   See individual terms: Validation and Verification

Standard Deviation of a Population   The Standard Deviation of an entire set (population) of data values is a measure of the extent to which the data is dispersed around its Arithmetic Mean. It is calculated as the square root of the Variance, which is the average of the squares of the deviations of each individual value from the Arithmetic Mean of all the values.

Standard Deviation of a Sample   The Standard Deviation of a sample of data taken from the entire population is a measure of the extent to which the sample data is dispersed around its Arithmetic Mean. It is calculated as the square root of the Sample Variance, which is the sum of squares of the deviations of each individual value from the Arithmetic Mean of all the values divided by the degrees of freedom, which is one less than the number of data points in the sample.

Standard Error   The Standard Error of a sample's statistic is the Standard Deviation of the sample values of that statistic around the true population value of that statistic. It can be approximated by the dividing the Sample Standard Deviation by the square root of the sample size.

Stanford-B Unit Learning Curve   A Stanford-B Unit Learning Curve is a variation of the Crawford Unit Learning Curve that allows for the benefits of prior learning to be expressed in terms of an adjustment to the effective number of cumulative units produced.

Statistics

(1)   The science or practice relating to the collection and interpretation of numerical and categorical data for the purposes of describing or inferring representative values of the whole data population from incomplete samples.

(2) The numerical values, measures and context that have been generated as outputs from the above practice.

Stepwise Regression Stepwise Regression by Forward Selection is a procedure by which a Multi-Linear Regression is compiled from a list of independent candidate variables, commencing with the most statistically significant individual variable (from a Simple Linear Regression perspective) and progressively adding the next most significant independent variable, until such time that the addition of further candidate variables does not improve the fit of the model to the data in accordance with the accepted Measures of Goodness of Fit for the Regression.

Stepwise Regression by Backward Elimination is a procedure by which a Multi-Linear Regression is compiled commencing with all potential independent candidate variables and eliminating the least statistically significant variable progressively (one at a time) until such time that all remaining candidate variables are deemed to be statistically significant in accordance with the accepted Measures of Goodness of Fit.

Subject Matter Expert's Opinion (Expert Judgement) Expert Judgement is a recognised term expressing the opinion of a Subject Matter Expert (SME).

SV (Schedule Variance – Cost Impact) See Earned Value Management Abbreviations and Terminology.

SV(t) (Schedule Variance – Time Impact) See Earned Value Management Abbreviations and Terminology.

Tertiary Data See Data Type.

Then Year Values 'Then Year Values' are values that have been adjusted to express an expectation of what might be incurred in the future due to escalation or other predicted changes. In some cases, these may be referred to as 'Outturn Year Values'.

Three-Point Estimate See 3-Point Estimate.

Time Series Analysis Time Series Analysis is the procedure whereby a series of values obtained at successive time intervals is separated into its constituent elements that describe and calibrate a repeating pattern of behaviour over time in relation to an underlying trend.

An Additive/Subtractive Time Series Model is one in which the Predicted Value is a function of the forecast value attributable to the underlying Trend plus or minus adjustments for its relative Seasonal and Cyclical positions in time.

A Multiplicative Time Series Model is one in which the Predicted Value is a function of the forecast value attributable to the underlying Trend multiplied by appropriate Seasonal and Cyclical Factors.

Time-Based Learning Curve A Time-based Learning Curve is an empirical relationship which reflects that the time, effort or cost to produce an output from an activity decreases as the elapsed time since commencement of that activity increases.

Time-Constant Learning Curve A Time-Constant Learning Curve considers the output or yield per time period from an activity rather than the time or cost to produce a unit. The model assumes that the output increases due to learning, from an initial starting level, before flattening out asymptotically to a steady state level.

Time-Performance Learning Curve A Time-Performance Learning Curve is an empirical relationship that expresses the reduction in the average time or cost per unit produced per period as a power function of the cumulative number periods since production commenced.

Top-down Approach In a top-down approach to estimating, the estimator reviews the overall scope of work in order to identify the major elements of work and characteristics (drivers) that could be estimated separately from other elements. Typically, the estimator might consider a natural flow down through the Work Breakdown Structure (WBS), Product Breakdown Structure (PBS) or Service Breakdown Structure (SBS). The estimate scope may be broken down to different levels of WBS etc as required; it is not necessary to cover all elements of the task at the same level, but the overall project scope must be covered. The overall project estimate would be created by aggregating these high-level estimates. Lower level estimates can be created by subsequent iterations of the estimating process when more definition becomes available, and bridging back to the original estimate.

TRACEability A Basis of Estimate should satisfy the principles of TRACEability:

Transparent – clear and unambiguous with nothing hidden

Repeatable – allowing another estimator to reproduce the same results with the same information

Appropriate – it is justifiable and relevant in the context it is to be used

Credible – it is based on reality or a pragmatic reasoned argument that can be understood and is believable

Experientially-based – it can be underpinned by reference to recorded data (evidence) or prior confirmed experience

Transformation See Mathematical Transformation.

Trusted Source Estimating Method The Trusted Source Method of Estimating is one in which the Estimate Value is provided by a reputable, reliable or undisputed source. Typically, this might be used for low value cost elements. Where the cost element is for a more significant cost value, it would not be unreasonable to request the supporting Basis of Estimate, but this may not be forthcoming if the supporting technical information is considered to be proprietary in nature.

t-Test A t-Test is used for small sample sizes (< 30) to test probability of getting a sample's test statistic (often the Mean), if the equivalent population statistic has an assumed different value. It is also used to test whether two samples could be drawn from the same population.

Tukey's Fences A test for a single Outlier based on the Inter-Quartile Range of the data sample.

Type I Error A Type I Error is one in which we accept a hypothesis we should have rejected.

Type II Error A Type II Error is one in which we reject a hypothesis we should have accepted.

U-Test See Mann-Whitney U-Test.

Uncertainty Uncertainty is an expression of the lack of exactness around a variable, and is frequently quantified in terms of a range of potential values with an optimistic or lower end bound and a pessimistic or upper end bound.

Validation (Spreadsheet or Model) Validation is the process by which the assumptions and data used in a spreadsheet or model are checked for accuracy and appropriateness for their intended purpose.

See also Verification.

Variance of a Population The Variance of an entire set (population) of data values is a measure of the extent to which the data is dispersed around its Arithmetic Mean. It is calculated as the average of the squares of the deviations of each individual value from the Arithmetic Mean of all the values.

Variance of a Sample The Variance of a Sample of data taken from the entire population is a measure of the extent to which the sample data is dispersed around its Arithmetic Mean. It is calculated as the sum of squares of the deviations of each individual value from the Arithmetic Mean of all the values divided by the degrees of freedom, which is one less than the number of data points in the sample.

Verification (Spreadsheet or Model) Verification is the process by which the calculations and logic of a spreadsheet or model are checked for accuracy and appropriateness for their intended purpose.

See also Validation.

Wilcoxon-Mann-Whitney U-Test See Mann-Whitney U-Test.

Wright's Cumulative Average Learning Curve Wright's Cumulative Average Learning Curve is an empirical relationship that expresses the reduction in the cumulative average time or cost of each unit produced as a power function of the cumulative number units produced.

Z-Score A Z-Score is a statistic which standardises the measurement of the distance of a data point from the Population Mean by dividing by the Population Standard Deviation.

Z-Test A Z-Test is used for large sample sizes (< 30) to test probability of getting a sample's test statistic (often the Mean), if the equivalent population statistic has an assumed different value.

Legend for Microsoft Excel Worked Example Tables in Greyscale

Cell type	Potential Good Practice Spreadsheet Modelling Colour	Greyscale used in Book	Example of Greyscale Used in Book
Header or Label	Light Grey	Text on grey	Text
Constant	Deep blue	Bold white numeric on black	1
Input	Pale Yellow	Normal black numeric on pale grey	23
Calculation	Pale Green	Normal black numeric on mid grey	45
Solver variable	Lavender	Bold white numeric on mid grey	67
Array formula	Bright Green	Bold white numeric on dark grey	89
Random Number	Pink	Bold black numeric on dark grey	0.0902
Comment	White	Text on white	Text

Index